The Charlotte Mason Elementary Arithmetic Series

Book 2

by Richele Baburina

Special thanks to Kristen Orr for her help in writing the sums.

The Charlotte Mason Elementary Arithmetic Series, Book 2

Cover Design: John Neiner
Cover Wrap Photo: Corin Jones
Book Design: Sarah Shafer and John Shafer

ISBN 978-1-61634-433-7 printed
ISBN 978-1-61634-434-4 electronic download

Published by
Simply Charlotte Mason, LLC
930 New Hope Road #11-892
Lawrenceville, Georgia 30045
simplycharlottemason.com

Printed in the USA

"The Principality of *Mathematics* is a mountainous land, but the air is very fine and health-giving, though some people find it too rare for their breathing. It differs from most mountainous countries in this, that you cannot lose your way, and that every step taken is on firm ground. People who seek their work or play in this principality find themselves braced by effort and satisfied with truth"

(*Ourselves*, Book 1, p. 38).

Contents

Introduction

Charlotte Mason desired that children not merely have a mechanical ability to work arithmetic problems but that they form a relationship with numbers. In fact, Irene Stephens, Ambleside's lecturer in mathematics—who would later become the director of Charlotte Mason's Parents' Union School—tells us that Miss Mason referred to arithmetic as "Number" in these initial years because she "wanted the children to get a real sense of number; some vision of its innate power and beauty far beyond the sum of the moment" (*Parents' Review,* Vol. XL, 1929, "Number: A Figure and a Step Onward," p. 35).

Charlotte Mason called education an *atmosphere*, a *discipline*, and a *life* and, likewise, you will see a number of methods distinct to her principles and applied philosophy of education throughout. Some to keep in mind:

Atmosphere. Lessons are short, 20 minutes in all, which includes a scheduled five minutes of rapid oral work. The teaching is carefully graduated with the child answering small word problems within his understanding while also providing mental effort. Lessons remain mainly oral, and children work out the tables with the use of concrete objects — what we today refer to as *manipulatives*. These objects are used to help the child understand the *why* of the work done. As the facts are proven, when a level of comfort is reached, and before boredom sets in, the everyday objects are put away.

Discipline. The methods employed in the arithmetic lessons are the means by which good habits are thoughtfully cultivated in the child. Charlotte Mason tells us that if we let the "arithmetic lesson be to the child a daily exercise in clear thinking and rapid, careful execution," the child's "mental growth will be as obvious as the sprouting of seedlings in the spring" (*Home Education,* p. 261). Both the variety of the questions and the liveliness with which they are given help develop a child's power of imagination and readiness and attention to his work. Though lessons are largely oral, the opportunity for more writing in Book 2 means neatness and careful execution should be adhered to. We never want the mechanics of writing to outweigh the ideas being presented in arithmetic, so the amount of written work assigned is largely dependent upon the child's ability. Having your child speak aloud as he solves larger problems and requiring fully worded answers during table work is another means used in training good habits. (For more on habit training and mathematics see the handbook *Mathematics: An Instrument of Living Teaching* by Richele Baburina.)

Life. The unfolding of ideas in a child's mind is a hallmark of a Charlotte Mason education. In this book, as the language of mathematics is presented to the child, a number of ideas will be met with; such as subtraction as the complement of addition, multiplication as the extension of addition, and the relationship between division and subtraction, to name a few. Simple, everyday objects serve as tools to see the rationale of the tables and, since the objects don't require too much teaching, never replace the importance of the idea being presented. The lessons are straightforward—without rule after rule or lengthy explanations—in order to allow the child to think and discover for himself.

As you guide your child through these lessons, we hope he experiences the beauty and enjoyment to be found in Number as the realm of mathematics continues to open to him.

Arithmetic Concepts in Book 2

- The idea and operation of addition
- The idea and operation of subtraction
- Addition and subtraction tables up to 20 (10 + 10)
- Addition and subtraction with money
- Addition and subtraction in interesting story problems
- Addition and subtraction with pure number
- Longer sums with three or four sets of figures
- Addition and subtraction of numbers within 1000
- The idea and operation of multiplication
- Simple multiplication to show the idea of "times"
- The symbol "x"
- Construction of multiplication tables up to 6 x 10
- Mastery of multiplication tables up to 6 x 10
- Multiplication by 10s, 100s, 1000s
- The idea and operation of division
- The symbol "÷"
- Multiplication and division with money
- Multiplication and division in interesting story problems
- Multiplication with pure number
- Short division with pure number
- Numeration and Notation of numbers up to 1000

Overview of Lessons

In Book 1, the student thoroughly investigated numbers up to 100 while establishing a solid foundation in place value. In this second book, he will receive a formal introduction to the four rules (addition, subtraction, multiplication, division) and tables; master addition and subtraction tables up to 20 (10 + 10) and multiplication tables up to 6 x 10; embark on division; and learn to read, write, and work with numbers up to 1,000.

Tables

Tables must be learned, but Charlotte Mason's approach helps the teacher prepare the child for this task in the best way possible. The tables are committed to memory in a variety of ways through work in the concrete, which progresses to imagined objects, then advances to pure number. The idea of subtraction as the complement of addition is reinforced as subtraction tables are worked out alongside addition tables. A child further experiences a natural overlap with multiplication as it is introduced as repeated addition through simple and interesting problems. Charlotte Mason's multiplication tables are distinctive, and their construction—either with objects or written—allows a child to grasp their rationale, discover patterns, and take division as a normal counterpart when ascertaining how many times a certain number is contained in another.

Tables built in the concrete by your child may easily be saved for the next day's work if constructed on a tray. A sheet of felt laid on the tray makes a good cushion under the items to prevent their moving or sliding between lessons. If you are not able to save the table, you may quickly build it in advance of the lesson time in order to save time on the following day.

If your child has advanced past the need to use a concrete table, by all means put the objects away and continue the oral questions without the aid of the manipulative; but feel the freedom to get them back out if need be. If you are working with an older child who needs to solidify his knowledge of the tables, and he has established an ease with handwriting, he may write the tables in his math notebook instead of building them with objects.

Mental Arithmetic

A distinguishing characteristic of Charlotte Mason Arithmetic is the five minutes of scheduled mental arithmetic each day. Since arithmetic lessons are mainly oral and require mental work, this activity is somewhat set apart by giving questions in a rapid, lively fashion to fix attention while promoting a child's promptness and concentration. In Book 2, the Mental Arithmetic time is most often taken with rapid table work but also includes questions that may involve more than one operation as well as numeration questions. Many of the questions given in the sections *Table Work* and *Review* work well for this time of Mental Arithmetic. Additional questions are found in the back of this book. Allow your child to make up questions of his own, or if you have multiple students, they may take turns posing questions to one another. This activity need not be relegated to the schoolroom alone and can easily take place out of doors, in the car, or around the table.

When to Advance

No book can give the exact number of equations needed for each individual child to solidify his understanding. As you work with your child, daily drawing out new power, you will want to progress carefully enough that he has attained a level of comfort before moving on. You should also feel the freedom to not use every question given if it is not necessary. The questions are written in such a way as to allow you to easily change numbers, names, and objects that relate to your own child's life and interests. A brief look through the lesson ahead of time will aid your ability to keep the lesson engaging by seeing where you are headed. Be sure to follow the lesson with something that uses a different part of the brain or body, such as singing, handwriting, handicraft, dance, or drill.

Pacing

While the concepts found in Book 2 generally take place in a child's second year of formal education, you will want to progress at a pace that ensures each step is taken on solid ground. The charts on the next page are given as an example and are merely a loose planning guide for your convenience. You may find that your child moves a bit more quickly through certain sections and a bit more slowly through others. That is perfectly normal. Consistency in daily arithmetic lessons is an important key to progress, but, as always, it is the chart that should be adjusted to your child's pace rather than trying to make your child fit the charts.

Term 1												
Suggested Weeks	1	2	3	4	5	6	7	8	9	10	11	12
Review of Previous Year's Work	■	■	■									
Addition and Subtraction Tables				■	■	■	■	■	■	■	■	
Term Exam												■

Term 2												
Suggested Weeks	13	14	15	16	17	18	19	20	21	22	23	24
Addition with Larger Numbers	■	■	■	■	■							
Subtraction with Larger Numbers						■	■	■	■			
Multiplication										■	■	
Term Exam												■

Term 3												
Suggested Weeks	25	26	27	28	29	30	31	32	33	34	35	36
Multiplication (cont'd).	■	■	■	■	■	■	■					
Division								■	■	■		
Review of Year's Work											■	
Term Exam												■

Exams

Three end-of-term exams are provided in the back of this book. Those oral exams are designed for you to customize to reflect where your child is in the lessons. See page 315 for details.

Supplies Needed

- Personal chalkboard, dry-erase board, or the like together with its appropriate writing instrument. For brevity, it will be referred to as *the slate* in the lessons. *
- Variety of everyday objects: beans, pennies, simple counters, candy-coated chocolates, beads, buttons, popcorn kernels, etc.
- Objects bundled in 10s from the previous year's work or a variety of objects that can either be bundled or threaded in groups of 10s. Beads or buttons can be threaded with string or chenille stems; craft sticks, matchsticks, or golf pencils are easily bundled with elastic bands. *
- Coin purse with 50 pennies, 5 nickels, 20 dimes, and 4 quarters; plus 100 pennies for building tables.
- 20 one-dollar bills, either real or play. *
- Gridded math notebook (sized according to child's handwriting: ¾", ½", or ¼"). *
- Pencil.
- Timer to be set for 20-minute lessons (15 minutes for the main portion and 5 minutes for mental arithmetic) or a clock to keep an eye on the time.

Optional

- The parent reference book, *Mathematics: An Instrument for Living Teaching,* explains Charlotte Mason's unique approach in mathematics more fully, from an introduction to numbers through geometry. *
- The video series, *Charlotte Mason's Living Math: A Guided Journey,* demonstrates the methods used in these lessons. *
- It is good to have on hand a collection of cards with clearly printed number sentences for convenience. The cards should include both a combination of numbers already studied as well as the various compositions of the number currently being taken. Before the

lesson begins, pull out an applicable card or two. This way, if you unexpectedly need to attend to something else for a few minutes, your child may continue working on his own without interruption. He may complete the number sentences, with manipulatives if needed, writing the answers on his slate. Preprinted Number Sentence Cards are provided in *The Charlotte Mason Elementary Arithmetic Series, Kit 2.* *

* *All items marked with an asterisk are available from Simply Charlotte Mason.*

Review from Book 1

Review at the beginning of a new school year is normal and, oftentimes, necessary. So remain at ease if it takes a bit of time for your child to do exercises he accomplished with little effort at the end of the previous school year. In fact, J. S. Mills tells us in a *Parents' Review* article regarding school vacation that he is "not at all afraid of retrogression," asserting that the qualities of a child's mind are such that a break of eight weeks is "sufficient for him to forget only very superficially the lessons of a term" and much prefers children to use the break to develop their senses, observation, and experience of life (*Parents' Review,* Vol. IV, 1893/94, p. 510).

You will find in this section a review of concepts covered in Book 1 taught in a Charlotte Mason way. Use it to refresh your student's memory of the ideas he discovered last year.

Tip: If you are continuing straight from Book 1 without a long break, this review section may be skipped.

You may also use this section as a placement guide to help you determine where your student should start in *The Charlotte Mason Elementary Arithmetic Series.* If your child has never worked with these concepts—or you believe he needs additional review—please see Book 1 for completely guided lessons. Book 2 gives a formal introduction to some but not all of the ideas explored in Book 1, along with a number of new concepts. See the section *Arithmetic Concepts in Book 2* on page 9 for details.

Tip: When determining placement, try to ascertain if your child has an understanding of the ideas contained in each section or merely possesses a mechanical ability to work the problems. One way to tell is whether his mathematical thinking is isolated to the arithmetic lesson only or is evident in his daily life.

The Symbol for the Idea *(instructions to the parent)*

1. Have your child point out *three* of something in the room, e.g., three apples, three books, three cups, three pencils, etc.

2. Have your child write the number *3* on his slate, getting it as neat as possible.

3. Repeat the two steps above with a few more numbers under ten (such as 6, 2, 5, 1, and 7), having your child first find the appropriate number of objects and then write the number as neatly as possible on his slate.

Simple Sums *(to be worked orally with or without manipulatives)*

Objects suggested: pennies

1. Abby bought a gum ball for 2¢ and a candy for 3¢. How much did she give for both?

Tip: Be sure to require fully worded answers; i.e., "2 and 3 are 5" or "2 plus 3 equals 5" for the answer above.

2. Hudson has 4¢ in his hand and 4¢ in his pocket. How much has he in all?

3. If you had 9¢ and lost 5¢, how much have you left?

4. Lyla has 5¢ and Macy has 5¢. How much have they all together?

5. Gabe had 6 pennies and Luke gave him 3 more pennies. How many pennies does Gabe have now?

6. Silas dug 4 potatoes and his father dug 6 more. How many potatoes do they have all together?

7. 3 geese were on a pond, then 2 more geese joined them. How many geese in all?

8. Jonah had 3 red balloons and 3 blue balloons. How many balloons had he in all?

9. There were 9 bunnies in a field when 2 hopped away. How many bunnies remained?

10. Annalise wants 8 flowers for her bouquet. She has picked 5. How many more does she need?
    - Do you remember how to write this equation using the signs you have learned? *(Child writes on slate either 5 + 3 = 8 or 8 – 5 = 3. Either is correct.)*
    - And how is it read?

Answers

1. 5¢, because 2 + 3 = 5

Tip: Be sure to require fully worded answers.

2. 8¢
3. 4¢
4. 10¢
5. 9 pennies
6. 10 potatoes
7. 5 geese
8. 6 balloons
9. 7 bunnies
10. 3 flowers; 5 + 3 = 8 or 8 – 5 = 3

Counting

1. Count out 10 pennies. Now count them forward and back.

2. Can you arrange them in groups of two? Now count them by 2s.

3. Now arrange them in groups of five. Can you count them by 5s?

Exchanging Coins *(to be worked orally with coins)*

Objects used: coin bag with quarters, dimes, nickels, and pennies

Tip: Have your student always keep dimes on his left and pennies on his right.

1. Count out 12 pennies. If you don't want to carry so many coins, what can you do? How many pennies have you left?

2. If you have 5 pennies and 1 nickel, what coin can you exchange them for?

3. You are a shopkeeper and must give me 23¢ in change. If I want as few coins as possible, what might you give me?

4. How many nickels can you change for ten pennies?

5. How many nickels can you change for a dime?

6. If we went to the store and you wanted to buy a candy that cost 25¢, which one coin could you give the cashier?

7. What other ways can you make 25¢?

8. Count out $1.00 from your coin bag using dimes. Count by 10s to 100.

9. You are a shopkeeper. I purchase a piece of gum from you for 25¢ and give you 3 dimes, or 30¢, in payment. How much change should you give me in return?

10. Elinor purchased a bookmark for 35¢. If she gave the shopkeeper 40¢, how much should she receive in change? And how would you write this equation? *(Child writes on slate.)*

11. If you have 2 quarters, how much have you?

12. If you changed your 50¢ into dimes, how many dimes would you have?

13. Aviya purchased a lemon for 39¢. He gave the shopkeeper 50¢. How much change should he receive in return?

14. You need 60¢ to purchase a colored pencil. You have 55¢. How much more do you need?

15. Can you show me two different ways to make 75¢?

16. Wyatt purchased an ice cream for 68¢. If he gave the shopkeeper 75¢, how much should he receive in change?

17. If you have 75¢, how many 10¢ stickers can you buy?

18. Aurav bought a pencil for 75¢ and an eraser for 25¢. How much did he give for both?

19. Show me 93¢ in dimes and pennies.

20. If you want to exchange pennies for a $1 bill, how many pennies do you need? And if you want to exchange your $1 bill for dimes, how many dimes will you receive?

Answers

1. Exchange 10 pennies for 1 dime; 2 pennies left

Tip: Be sure to require fully worded answers.

2. 1 dime
3. 2 dimes and 3 pennies
4. 2 nickels
5. 2 nickels
6. A quarter, or 25¢ piece
7. 25 pennies, 5 nickels, or 2 dimes and a nickel (or any other number of combinations the child is able to give)
8. 10 dimes counted 10, 20, 30, 40, 50, 60, 70, 80, 90, 100
9. 5¢ (in either a nickel or 5 pennies), because 30 – 25 = 5
10. 5¢; 35 + 5 = 40 or 40 – 35 = 5

11. 50¢
12. 5 dimes
13. 11¢, because 50 – 39 = 11 or 39 + 11 = 50
14. 5¢, because 55 + 5 = 60 or 60 – 55 = 5
15. 3 quarters, 7 dimes and a nickel (or any other various combinations the child is able to make correctly)
16. 7¢, because 75 – 68 = 7 or 68 + 7 = 75
17. 7 stickers
18. $1.00, because 75 + 25 = 100
19. 9 dimes and 3 pennies
20. 100 pennies; 10 dimes

Numeration and Notation *(instructions to the parent)*

Objects used: coins

1. Place 46¢ on the table using dimes and pennies. Be sure to keep proper place value by putting 4 dimes on the student's left and 6 pennies on the student's right. Ask: How much money do we have here?

2. Ask: How many units have we?

3. Next ask: How many tens have we?

4. Write *46* on the slate and ask something like: In numbers such as *46*, which figure represents units and which represents tens?

5. Take away the pennies and place another dime in line with the other 4 dimes to make 5 dimes on the student's left. Ask: Now how many tens have we? And how many units?

6. Write the following numbers in a column on either the slate or on grid paper and ask your child to read them: *4, 12, 25, 63, 82, 100.*

7. Now point to each number and ask how many units, tens, or hundreds in each.

8. Ask: In the number *63,* how do we know it is read "sixty-three" and not "thirty-six"?

9. Ask your child to write the following numbers in a column either on the slate or in his math notebook, being sure to keep proper place value, while you dictate: *2, 38, 17, 64, 29, 91, 100.*

Answers

1. 46¢
2. 6 units
3. 4 tens
4. The number to the far right, *6,* represents units and the number to the left of it, *4,* shows tens
5. 5 tens and 0 units
6. Four, twelve, twenty-five, sixty-three, eighty-two, one hundred
7. 4 units; 1 ten and 2 units; 2 tens and 5 units; 6 tens and 3 units; 8 tens and 2 units; 1 hundred, no tens, and no units
8. The *3* is in the units place and the *6* is in the tens place
9. Written in a column: *2, 38, 17, 64, 29, 91, 100*

Addition and Subtraction Tables

Addition and subtraction tables will simultaneously be worked out using concrete objects, with the idea being given that subtraction is the counterpart of addition. Since subtraction is usually the more difficult operation for a child, please be sure to advance as slowly as necessary for him to feel he is on solid ground. The learning of each table follows the same rhythm, laid out here as seven separate steps, with a review of previous tables learned.

It is important to note that children advance at different rates. If your child needs more sums in any given step, the samples provided can easily be changed to offer more practice. If you find your child is already thinking in numbers and has become bored or worried by the manipulatives, do not feel tied to them. Simply have him draw the table on his slate or write it in his math notebook for use during exercises, always knowing that concrete objects can be brought back out if needed.

Tip: The term sum *in Charlotte Mason's day referred to any mathematical problem. It will be used in that context in these printed lessons to the teacher. The modern term* equation *should be used when speaking to the student.*

To view demonstrations, watch the session, "Addition and Subtraction Tables," of the video series *Charlotte Mason's Living Math: A Guided Journey.*

Addition and Subtraction Table for 1

Step 1: Assemble a table and speak it through.

(instructions to the parent)

Objects used: pennies

1. Starting with the number 1, have your child assemble an addition table using his pennies. See sample diagram below. Place 1 penny with 0 pennies next to it. Under that, place 1 penny with 1 penny next to it. Then 1 penny with 2 pennies next to it, 1 penny with 3 pennies next to it, and so on up to 1 penny with 10 pennies next to it. Have your child say the table aloud as he arranges the pennies: 1 penny + 0 pennies = 1 penny; 1 penny + 1 penny = 2 pennies; 1 penny + 2 pennies = 3 pennies, etc.

Tip: Tables can be made on a piece of felt or on a tray in order to preserve work for the next day or, once the child has made the table on his own and worked with it for a lesson, it can be put away and, prior to the next day's work, you can quickly recreate the table in advance of the lesson to save time.

Tip: You could also allow the student to create the tables using dots or strokes on graph paper or on a gridded dry erase board, such as the one available through Simply Charlotte Mason. This option works well if your child balks at the idea of using concrete objects, if you are working with an older child to solidify his knowledge of the tables, or if concrete objects are no longer necessary.

Step 2: Pose simple addition sums using the objects in the table.

Tip: Skip around on the table; don't just go in order from top to bottom.

1. 1 penny + 3 pennies =
2. 1 penny + 7 pennies =
3. 1 penny + 1 penny =
4. 1 penny + 2 pennies =
5. 1 penny + 8 pennies =
6. 1 penny + 0 pennies =
7. 1 penny + 4 pennies =
8. 1 penny + 9 pennies =
9. 1 penny + 6 pennies =
10. 1 penny + 5 pennies =

Answers

1. 4 pennies

   Tip: Always encourage a full answer; e.g., 1 penny and 3 pennies make 4 pennies or 1 penny plus 3 pennies equal 4 pennies.

2. 8 pennies
3. 2 pennies
4. 3 pennies
5. 9 pennies
6. 1 penny
7. 5 pennies

8. 10 pennies
9. 7 pennies
10. 6 pennies

Tip: Remember to keep lessons only 15 minutes long, then spend 5 minutes in review. Next time, pick up where you left off.

Step 3: Progress to addition sums with imaginary objects.

Tip: Incorporating a variety of imaginary objects will help your child see that the addition facts are general and not tied to the specific manipulative we are using at the moment.

Tip: Remember to work at your child's pace. If he needs to, let him use the pennies to obtain the answer.

1. 1 kitten + 2 kittens = how many kittens?
2. 1 puppy + 6 puppies = how many puppies?
3. 1 pencil + 9 pencils = how many pencils?
4. 1 boat + 4 boats = how many boats?
5. 1 rabbit + 1 rabbit = how many rabbits?
6. 1 apple + 10 apples = how many apples?
7. 1 gum ball + 5 gum balls = how many gum balls?
8. 1 snowball + 7 snowballs = how many snowballs?
9. 1 snowman + 8 snowmen = how many snowmen?
10. 1 lamb + 3 lambs = how many lambs?

Answers

1. 3 kittens

Tip: Always encourage a full answer; e.g., 1 kitten and 2 kittens make 3 kittens or 1 kitten plus 2 kittens equal 3 kittens or 1 and 2 make 3.

2. 7 puppies
3. 10 pencils
4. 5 boats
5. 2 rabbits
6. 11 apples
7. 6 gum balls
8. 8 snowballs
9. 9 snowmen
10. 4 lambs

Tip: When the answers are coming readily to the sums with imaginary objects, your child may begin simple sums with pure numbers.

Step 4: Work with addition sums using pure number.

1. 1 + 4 =

2. 1 + 7 =

3. 1 + 3 =

4. 1 + 0 =

5. 1 + 5 =

6. 1 + 2 =

7. 1 + 9 =

8. 1 + 10 =

9. 1 + 8 =

10. 1 + 1 =

Answers

1. 5

Tip: Be sure to require fully worded answers: 1 + 4 = 5.

2. 8
3. 4
4. 1
5. 6
6. 3
7. 10
8. 11
9. 9
10. 2

Step 5: Incorporate the subtraction table as addition continues.

Go over each line in the addition table you have built and have your child work out both the addition and the complementary subtraction facts. To begin with, word the subtraction questions in this way: You have 1 penny but want two pennies. How many more must you get? This kind of wording reinforces the idea of subtraction as the complement of addition and may be easier for the student to grasp. You may eventually proceed to word the subtraction questions as, 3 pennies take away 1 penny leaves how many pennies?

Tip: Feel free to mix up the order of the problems as your child becomes more comfortable in the work.

1. 1 penny added to 1 penny make how many pennies?

2. You have 1 penny but you want 2 pennies. How many more must you get?

3. 1 penny added to 2 pennies make how many pennies?

4. You have 1 penny but want 3 pennies. How many more must you get?

5. 1 penny added to 3 pennies make how many pennies?

6. You have 1 penny and you want 4 pennies. How many more must you get?

7. 1 penny added to 4 pennies make how many pennies?

8. You have 1 penny and want 5 pennies. How many more must you get?

9. 1 penny added to 5 pennies make how many pennies?

10. You have 1 penny but want 6 pennies. How many more pennies do you need?

11. 1 penny added to 6 pennies make how many pennies?

12. 7 pennies take away 1 penny leaves how many pennies?

13. 1 penny added to 7 pennies equals how many pennies?

14. You have 8 pennies and take 1 away. How many pennies have you?

15. 1 penny added to 8 pennies make how many pennies?

16. You have 1 penny but want 9 pennies. How many more must you get?

17. 1 penny added to 9 pennies make how many pennies?

18. 10 pennies take away 1 penny leaves how many pennies?

19. 1 penny and 10 pennies together make how many pennies?

20. 11 pennies take away 1 penny leaves you with how many pennies?

Answers

1. 2 pennies

Tip: Be sure to require fully worded answers.

2. 1 penny
3. 3 pennies
4. 2 pennies
5. 4 pennies
6. 3 pennies
7. 5 pennies
8. 4 pennies
9. 6 pennies
10. 5 pennies
11. 7 pennies
12. 6 pennies
13. 8 pennies
14. 7 pennies
15. 9 pennies
16. 8 pennies
17. 10 pennies
18. 9 pennies
19. 11 pennies
20. 10 pennies

Step 6: Give mixed sums (both addition and subtraction) with imaginary objects.

Tip: On days when the work is going exceedingly well, you may have your child write on his slate or in his math notebook an addition or subtraction equation that he has solved, using the proper signs. If writing in the notebook, make sure each numeral and each symbol is written in its own square on the grid paper. At this point, do not assign sums worked out solely on paper. The work remains mostly oral.

1. 1 apple from 4 apples?
2. 1 cookie from 10 cookies?
3. 1 marble and 8 marbles?
4. 1 bird and 6 birds?
5. 1 fox from 7 foxes?
6. 1 frog from 3 frogs?
7. 1 owl plus 5 owls?
8. 1 puppy from 2 puppies?
9. 1 book and 7 books?
10. 1 tree from 8 trees?

Answers

1. 3 apples

Tip: Be sure to require fully worded answers.

2. 9 cookies
3. 9 marbles
4. 7 birds
5. 6 foxes
6. 2 frogs
7. 6 owls
8. 1 puppy
9. 8 books
10. 7 trees

Tip: When answers using imaginary objects come easily, begin giving simple mixed problems using pure number.

Step 7: Present mixed sums using pure number.

Tip: If your child is working comfortably with pure numbers and able to mentally imagine them, then you may put the manipulatives away until the next number is introduced.

How much is left if you take

1. 1 from 2?
2. 1 from 5?
3. 1 from 9?
4. 1 from 6?
5. 1 from 4?

How much have you if you add

6. 1 and 8?
7. 1 and 3?
8. 1 and 7?
9. 1 and 2?
10. 1 and 10?

Answers

Tip: Be sure to require fully worded answers.

1. 1
2. 4
3. 8
4. 5
5. 3
6. 9
7. 4
8. 8
9. 3
10. 11

Addition and Subtraction Table for 2

Tip: As a help, pull a Number Sentence Card out before the lesson begins so your student's work can continue uninterrupted if you get called away for a moment. (See page 13 for details.)

Step 1: Assemble a table and speak it through.

(instructions to the parent)

Objects used: pennies

1. Have your child assemble an addition table using his pennies, placing 2 pennies next to 0 pennies, 2 pennies next to 1 penny, 2 pennies next to 2 pennies, and so on up to 2 and 10. The child should say it through as he arranges the pennies: 2 pennies + 0 = 2 pennies; 2 pennies + 1 penny = 3 pennies; 2 pennies + 2 pennies = 4 pennies; etc.

••
•• •
•• ••
•• •••
•• ••••
•• •••••
•• ••••••
•• •••••••
•• ••••••••
•• •••••••••
•• ••••••••••

Tip: You could also allow the student to create the tables using dots or strokes on graph paper or on a gridded dry erase board. This option works well if your child balks at the idea of using concrete objects, if you are working with an older child to solidify his knowledge of the tables, or if concrete objects are no longer necessary.

Step 2: Pose simple addition sums using the objects in the table.

1. 2 pennies + 5 pennies =

2. 2 pennies + 8 pennies =

3. 2 pennies + 2 pennies =

4. 2 pennies + 7 pennies =

5. 2 pennies + 6 pennies =

6. 2 pennies + 3 pennies =

7. 2 pennies + 9 pennies =

8. 2 pennies + 10 pennies =

9. 2 pennies + 0 pennies =

10. 2 pennies + 4 pennies =

11. 2 pennies + 1 penny =

Answers

1. 7 pennies

Tip: Be sure to require fully worded answers: 2 + 5 = 7.

2. 10 pennies
3. 4 pennies
4. 9 pennies
5. 8 pennies
6. 5 pennies
7. 11 pennies
8. 12 pennies
9. 2 pennies
10. 6 pennies
11. 3 pennies

Step 3: Progress to addition sums with imaginary objects.

1. 2 stickers + 8 stickers = how many stickers?

2. 2 rings + 2 rings = how many rings?

3. 2 boats + 7 boats = how many boats?

4. 2 bikes + 4 bikes = how many bikes?

5. 2 erasers + 5 erasers = how many erasers?

6. 2 pencils + 1 pencil = how many pencils?

7. 2 oranges + 6 oranges = how many oranges?

8. 2 glue sticks + 3 glue sticks = how many glue sticks?

9. 2 necklaces + 10 necklaces = how many necklaces?

10. 2 bugs + 9 bugs = how many bugs?

Answers

1. 10 stickers

Tip: Be sure to require fully worded answers.

2. 4 rings
3. 9 boats
4. 6 bikes
5. 7 erasers
6. 3 pencils
7. 8 oranges
8. 5 glue sticks
9. 12 necklaces
10. 11 bugs

Tip: Move on to the next step only when answers are coming readily in the current step.

Step 4: Work with addition sums using pure number.

1. 2 + 5 =
2. 2 + 8 =
3. 2 + 4 =
4. 2 + 7 =
5. 2 + 1 =
6. 2 + 6 =
7. 2 + 3 =
8. 2 + 10 =
9. 2 + 2 =
10. 2 + 9 =

Answers

1. 7

Tip: Be sure to require fully worded answers.

2. 10
3. 6
4. 9
5. 3
6. 8
7. 5
8. 12
9. 4
10. 11

Step 5: Incorporate the subtraction table as addition continues.

Tip: Mix up the order of the problems as your child becomes more comfortable in the work.

1. 2 pennies added to 2 pennies equals how many pennies?
2. You have 7 pennies and take 2 away. How many pennies do you have?
3. 2 pennies added to 7 pennies make how many pennies?
4. You have 2 pennies but want 10 pennies. How many must you get?
5. 2 pennies added to 4 pennies make how many pennies?
6. You have 2 pennies and want 7 pennies. How many more must you get?
7. 2 pennies added to 9 pennies make how many pennies?
8. You have 2 pennies but want 3 pennies. How many more pennies do you need?
9. 8 pennies added to 2 pennies make how many pennies?
10. 2 pennies added to 1 penny make how many pennies?
11. You have 2 pennies but you want 4 pennies. How many more must you get?
12. 2 pennies added to 5 pennies make how many pennies?
13. You have 2 pennies but want 9 pennies. How many more must you get?

14. 2 pennies added to 3 pennies make how many pennies?

15. You have 2 pennies and you want 5 pennies. How many more must you get?

16. 6 pennies added to 2 pennies make how many pennies?

17. 6 pennies take away 2 pennies leaves how many pennies?

18. 8 pennies take away 2 pennies leaves how many pennies?

19. 2 pennies and 10 pennies together make how many pennies?

20. 5 pennies take away 2 pennies leaves you with how many pennies?

Answers

1. 4 pennies

Tip: Be sure to require fully worded answers.

2. 5 pennies
3. 9 pennies
4. 8 pennies
5. 6 pennies
6. 5 pennies
7. 11 pennies
8. 1 penny
9. 10 pennies
10. 3 pennies
11. 2 pennies
12. 7 pennies
13. 7 pennies
14. 5 pennies
15. 3 pennies
16. 8 pennies
17. 4 pennies
18. 6 pennies

19. 12 pennies
20. 3 pennies

Step 6: Give mixed sums with imaginary objects.

Tip: On days when the work is going exceedingly well, you may have your child write on his slate or in his math notebook an addition or subtraction equation that he has solved, using the proper signs.

1. 2 birds from 5 birds?

2. 2 rocks and 4 rocks?

3. 2 leaves plus 10 leaves?

4. 2 eggs from 4 eggs?

5. 2 cookies and 7 cookies?

6. 2 seashells and 5 seashells?

7. 2 birds from 3 birds?

8. 2 cards added to 2 cards?

9. 2 marbles plus 8 marbles?

10. 2 candles from 12 candles?

Answers

1. 3 birds

Tip: Be sure to require fully worded answers.

2. 6 rocks
3. 12 leaves
4. 2 eggs

5. 9 cookies
6. 7 seashells
7. 1 bird
8. 4 cards
9. 10 marbles
10. 10 candles

Step 7: Present mixed sums using pure number.

Tip: If the child is no longer needing to look at the objects, put the manipulatives away and continue the work without their use until the next number is introduced.

How much is left if you take

1. 2 from 3?
2. 2 from 8?
3. 2 from 2?
4. 2 from 9?
5. 2 from 5?

How much have you if you add

6. 2 and 9?
7. 2 and 6?
8. 2 and 1?
9. 2 and 5?
10. 2 and 3?

Answers

Tip: Be sure to require fully worded answers.

1. 1
2. 6
3. 0
4. 7
5. 3
6. 11
7. 8
8. 3
9. 7
10. 5

Addition and Subtraction Table for 3

Tip: Keep lessons to 15 minutes of work on the newest concept and 5 minutes of review at the end, making each day's math time 20 minutes total.

Step 1: Assemble a table and speak it through.

Objects used: beads

```
•••
•••     •
•••     ••
•••     •••
•••     ••••
•••     •••••
•••     ••••••
•••     •••••••
•••     ••••••••
•••     •••••••••
•••     ••••••••••
```

Tip: If you are working with an older child to solidify his knowledge of the tables, and handwriting comes easily, the child may write the table in his math notebook rather than use objects to build it.

Step 2: Pose simple addition sums using the objects in the table.

Tip: After the initial day of building the new table, if you are not able to save it, then you may quickly build the table yourself before the lesson begins.

1. 3 beads + 7 beads =

2. 3 beads + 5 beads =

3. 3 beads + 8 beads =

4. 3 beads + 2 beads =

5. 3 beads + 9 beads =

6. 3 beads + 6 beads =

7. 3 beads + 1 bead =

8. 3 beads + 4 beads =

9. 3 beads + 3 beads =

10. 3 beads + 0 beads =

11. 3 beads + 10 beads =

Answers

1. 10 beads

Tip: Be sure to require fully worded answers: 3 + 7 = 10.

2. 8 beads
3. 11 beads
4. 5 beads
5. 12 beads
6. 9 beads
7. 4 beads
8. 7 beads
9. 6 beads
10. 3 beads
11. 13 beads

Step 3: Progress to addition sums with imaginary objects.

1. 3 ribbons + 8 ribbons = how many ribbons?

2. 3 dogs + 4 dogs = how many dogs?

3. 3 stamps + 7 stamps = how many stamps?

4. 3 birds + 6 birds = how many birds?

5. 3 chairs + 2 chairs = how many chairs?

6. 3 gum balls + 10 gum balls = how many gum balls?

7. 3 cats + 5 cats = how many cats?

8. 3 rabbits + 1 rabbit = how many rabbits?

9. 3 cards + 9 cards = how many cards?

10. 3 kittens + 3 kittens = how many kittens?

Tip: Remember not to move on to the next step until your child is working comfortably in the current one.

Answers

1. 11 ribbons

Tip: Be sure to require fully worded answers.

2. 7 dogs
3. 10 stamps
4. 9 birds
5. 5 chairs
6. 13 gum balls
7. 8 cats
8. 4 rabbits
9. 12 cards
10. 6 kittens

Step 4: Work with addition sums using pure number.

1. 3 + 5 =

2. 3 + 9 =

3. 3 + 6 =

4. 3 + 7 =

5. 3 + 3 =

6. 3 + 1 =

7. 3 + 0 =

8. 3 + 4 =

9. 3 + 8 =

10. 3 + 2 =

Answers

1. 8

Tip: Be sure to require fully worded answers.

2. 12
3. 9
4. 10
5. 6
6. 4
7. 3
8. 7
9. 11
10. 5

Step 5: Incorporate the subtraction table as addition continues.

1. 3 beads added to 2 beads make how many beads?

2. You have 3 beads but want 7 beads. How many must you get?

3. 3 beads added to 10 beads make how many beads?

4. 9 beads take away 3 beads leaves how many beads?

5. 3 beads and 9 beads together make how many beads?

6. 10 beads take away 3 beads leaves you with how many beads?

7. 3 beads added to 4 beads make how many beads?

8. You have 3 beads and want 9 beads. How many more must you get?

9. 3 beads added to 7 beads together make how many beads?

10. You have 3 beads but want 8 beads. How many more beads do you need?

11. 3 beads added to 6 beads make how many beads?

12. 5 beads take away 3 beads leaves how many beads?

13. 3 beads added to 8 beads equals how many beads?

14. You have 8 beads and take 3 away. How many beads do you have?

15. 3 beads added to 2 beads make how many beads?

16. You have 3 beads but you want 6 beads. How many more must you get?

17. 3 beads added to 5 beads make how many beads?

18. You have 3 beads but want 7 beads. How many more must you get?

19. 3 beads added to 3 beads make how many beads?

20. You have 3 beads and you want 4 beads. How many more must you get?

Answers

1. 5 beads

Tip: Be sure to require fully worded answers.

2. 4 beads
3. 13 beads
4. 6 beads
5. 12 beads
6. 7 beads
7. 7 beads
8. 6 beads
9. 10 beads
10. 5 beads
11. 9 beads
12. 2 beads
13. 11 beads
14. 5 beads
15. 5 beads
16. 3 beads
17. 8 beads
18. 4 beads
19. 6 beads
20. 1 bead

Step 6: Give mixed sums with imaginary objects.

Tip: On days when the work is going exceedingly well, you may have your child write on his slate or in his math notebook an addition or subtraction equation he has solved, using the proper signs.

1. 3 peaches from 6 peaches?

2. 3 plums from 5 plums?

3. 3 rubber bands and 4 rubber bands?

4. 3 bananas from 7 bananas?

5. 3 dinosaurs and 8 dinosaurs?

6. 3 puppies from 8 puppies?

7. 3 seeds and 9 seeds?

8. 3 sticks and 5 sticks?

9. 3 rings from 3 rings?

10. 3 lions and 7 lions?

Answers

1. 3 peaches

Tip: Be sure to require fully worded answers.

2. 2 plums
3. 7 rubber bands
4. 4 bananas
5. 11 dinosaurs
6. 5 puppies
7. 12 seeds
8. 8 sticks
9. 0 rings
10. 10 lions

Step 7: Present mixed sums using pure number.

Tip: Put the manipulatives away once your child has achieved a comfort working with this table mentally.

How much is left if you take

1. 3 from 5?
2. 3 from 6?
3. 3 from 9?
4. 3 from 4?
5. 3 from 13?

How much have you if you add

6. 3 and 9?
7. 3 and 4?
8. 3 and 8?
9. 3 and 10?
10. 3 and 5?

Answers

Tip: Be sure to require fully worded answers.

1. 2
2. 3
3. 6
4. 1
5. 10
6. 12
7. 7
8. 11
9. 13
10. 8

Review 3 and Previous Tables: Addition

1. 2 butterflies in the garden and 5 more join them. How many all together?
2. 2 hens laid eggs and later 4 more laid eggs. How many hens laid eggs that morning?
3. Kristen made 3 cookies on the first tray and 9 on the second. How many cookies did she make in all?
4. There is 1 fox in the field and 5 foxes in the woods. How many foxes in all?
5. The baby ate 1 jar of carrots and 1 jar of bananas. How many jars of food did she eat?

6. 1¢ + 4¢ =

7. 2¢ + 6¢ =

8. 3¢ + 5¢ =

9. 2¢ + 7¢ =

10. 3¢ + 2¢ =

11.

2 pigs	3 sheep	1 ball	3 wrens	1 block
+ 9 pigs	+ 3 sheep	+ 9 balls	+ 4 wrens	+ 7 blocks

12. 3 + 4

13. 1 + 7

14. 2 + 2

15. 2 + 8

16. 1 + 10

Answers

1. 7 butterflies

Tip: Be sure to require fully worded answers.

2. 6 hens
3. 12 cookies
4. 6 foxes
5. 2 jars
6. 5¢
7. 8¢
8. 8¢
9. 9¢
10. 5¢

11.

2 pigs	3 sheep	1 ball	3 wrens	1 block
+ 9 pigs	+ 3 sheep	+ 9 balls	+ 4 wrens	+ 7 blocks
11 pigs	6 sheep	10 balls	7 wrens	8 blocks

12. 7
13. 8
14. 4
15. 10
16. 11

Review 3 and Previous Tables: Subtraction

1. There are 7 dogs at the park. 3 leave with their owners. How many dogs are left?

2. 6 ponies are eating grass in the field. 2 trot away. How many ponies are still eating grass?

3. 5 birds are sitting on a giraffe when1 flies away. How many are left?

4. Benny had 5¢ and spent 2¢. How much did he have left?

5. 8 birds are sitting on the wire and 3 fly away. How many birds remain?

6. 3¢ – 1¢ =

7. 6¢ – 1¢ =

8. 1¢ – 0¢ =

9. 7¢ – 2¢ =

10. 2¢ – 2¢ =

11.

3 seashells	2 rulers	6 highlighters	5 blocks	4 ribbons
– 2 seashells	– 1 ruler	– 3 highlighters	– 3 blocks	– 2 ribbons

12. 3 – 2

13. 3 – 3

14. 1 – 1

15. 4 – 3

16. 2 – 0

Answers

1. 4 dogs

Tip: Be sure to require fully worded answers.

2. 4 ponies
3. 4 birds
4. 3¢
5. 5 birds
6. 2¢
7. 5¢
8. 1¢
9. 5¢
10. 0¢
11.

3 seashells	2 rulers	6 highlighters	5 blocks	4 ribbons
– 2 seashells	– 1 ruler	– 3 highlighters	– 3 blocks	– 2 ribbons
1 seashell	1 ruler	3 highlighters	2 blocks	2 ribbons

12. 1
13. 0
14. 0
15. 1
16. 2

If your child needs more practice to be comfortable with this table, you can find additional review sums for 3 and previous tables on page 269. Also look at previous lessons for many more sums that you can use as is or by changing names, numbers, or objects.

Addition and Subtraction Table for 4

Tip: As a help, pull a Number Sentence Card out before the lesson begins so your student's work can continue uninterrupted if you get called away for a moment.

Step 1: Assemble a table and speak it through.

Objects used: beads

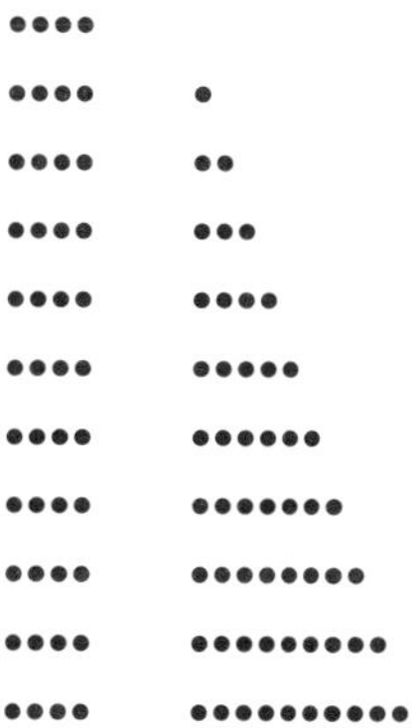

Tip: If you are working with an older child to solidify his knowledge of the tables, and handwriting comes easily, the child may write the table in his math notebook rather than use objects to build it.

Step 2: Pose simple addition sums using the objects in the table.

Tip: After the initial day of building the new table, if you are not able to save it, then you may quickly build the table yourself before the lesson begins.

1. 4 beads + 5 beads =

2. 4 beads + 8 beads =

3. 4 beads + 9 beads =

4. 4 beads + 3 beads =

5. 4 beads + 2 beads =

6. 4 beads + 4 beads =

7. 4 beads + 10 beads =

8. 4 beads + 6 beads =

9. 4 beads + 1 bead =

10. 4 beads + 7 beads =

11. 4 beads + 0 beads =

Answers

1. 9 beads

Tip: Be sure to require fully worded answers: 4 + 5 = 9.

2. 12 beads
3. 13 beads
4. 7 beads
5. 6 beads
6. 8 beads
7. 14 beads
8. 10 beads
9. 5 beads
10. 11 beads
11. 4 beads

Step 3: Progress to addition sums with imaginary objects.

1. 4 blocks + 4 blocks = how many blocks?

2. 4 buckets + 8 buckets = how many buckets?

3. 4 fish + 9 fish = how many fish?

4. 4 pears + 5 pears = how many pears?

5. 4 brushes + 7 brushes = how many brushes?

6. 4 socks + 2 socks = how many socks?

7. 4 coloring books + 3 coloring books = how many coloring books?

8. 4 rabbits + 6 rabbits = how many rabbits?

9. 4 crackers + 10 crackers = how many crackers?

10. 4 butterflies + 1 butterfly = how many butterflies?

Tip: Remember, don't move to the next step until answers are coming readily in the current step.

Answers

1. 8 blocks

Tip: Be sure to require fully worded answers.

2. 12 buckets
3. 13 fish
4. 9 pears
5. 11 brushes
6. 6 socks
7. 7 coloring books
8. 10 rabbits
9. 14 crackers
10. 5 butterflies

Step 4: Work with addition sums using pure number.

1. 4 + 5 =

2. 4 + 8 =

3. 4 + 7 =

4. 4 + 1 =

5. 4 + 3 =

6. 4 + 9 =

7. 4 + 2 =

8. 4 + 10 =

9. 4 + 4 =

10. 4 + 6 =

Answers

1. 9

Tip: Be sure to require fully worded answers.

2. 12
3. 11
4. 5
5. 7
6. 13
7. 6
8. 14
9. 8
10. 10

Step 5: Incorporate the subtraction table as addition continues.

Tip: Feel free to mix up the order of the problems as your child becomes more comfortable in the work.

1. 4 beads added to 6 beads make how many beads?
2. You have 4 beads but want 6 beads. How many more beads do you need?
3. You have 4 beads but want 5 beads. How many must you get?
4. 4 beads added to 9 beads make how many beads?
5. 12 beads take away 4 beads leaves how many beads?
6. 4 beads and 5 beads together make how many beads?
7. 9 beads take away 4 beads leaves you with how many beads?
8. 4 beads added to 10 beads make how many beads?
9. 5 beads take away 4 beads leaves how many beads?
10. 4 beads added to 3 beads equals how many beads?
11. You have 7 beads and take 4 away. How many beads do you have?
12. 4 beads added to 7 beads make how many beads?
13. 4 beads added to 8 beads make how many beads?
14. You have 4 beads but want 8 beads. How many more must you get?
15. 4 beads added to 2 beads make how many beads?

16. You have 4 beads and you want 9 beads. How many more must you get?

17. 4 beads added to 1 bead make how many beads?

18. 4 beads added to 4 beads make how many beads?

19. You have 4 beads but you want 7 beads. How many more must you get?

20. You have 4 beads and want 6 beads. How many more must you get?

Answers

1. 10 beads

Tip: Be sure to require fully worded answers.

2. 2 beads
3. 1 bead
4. 13 beads
5. 8 beads
6. 9 beads
7. 5 beads
8. 14 beads
9. 1 bead
10. 7 beads
11. 3 beads
12. 11 beads
13. 12 beads
14. 4 beads
15. 6 beads
16. 5 beads
17. 5 beads
18. 8 beads
19. 3 beads
20. 2 beads

Step 6: Give mixed sums with imaginary objects.

1. 4 keys from 9 keys?
2. 4 snails and 6 snails?
3. 4 leaves and 1 leaf?
4. 4 bugs from 5 bugs?
5. 4 trees from 8 trees?
6. 4 flowers from 7 flowers?
7. 4 frogs and 3 frogs?
8. 4 puppies from 12 puppies?
9. 4 books and 7 books?
10. 4 owls and 5 owls? *After solving, your student may write the equation 4 + 5 = 9 in his math notebook.*

Answers

1. 5 keys

Tip: Be sure to require fully worded answers.

2. 10 snails
3. 5 leaves
4. 1 bug
5. 4 trees
6. 3 flowers
7. 7 frogs
8. 8 puppies
9. 11 books
10. 9 owls

Step 7: Present mixed sums using pure number.

Tip: Be sure to put away the manipulatives when they are no longer needed for this table.

How much is left if you take

1. 4 from 8?
2. 4 from 9?
3. 4 from 5?
4. 4 from 7?
5. 4 from 10?

How much have you if you add

6. 4 and 8?
7. 4 and 6?
8. 4 and 2?
9. 4 and 5?
10. 4 and 9?

Answers

Tip: Be sure to require fully worded answers.

1. 4
2. 5
3. 1
4. 3
5. 6
6. 12
7. 10
8. 6
9. 9
10. 13

Review 4 and Previous Tables: Addition

1. Joshua ate 3 pieces of celery and his sister, Caroline, ate 2. How many pieces did they eat all together?
2. Robert gave 2 books to his sister and 6 books to his neighbor. How many books did he give away in all?
3. Mr. Gisel bought taffy for his children. He bought 4 pieces for Mary,

3 pieces for John, and 1 piece for Sarah. How many pieces of taffy did Mr. Gisel purchase?

4. Ralph bought 3 pounds of apples for applesauce and 5 pounds of apples to eat. How many pounds of apples did he purchase in all?

5. Elizabeth gathered 3 eggs on Friday and 6 eggs on Saturday. How many eggs did she gather all together?

6. 4¢ + 6¢ =

7. 2¢ + 8¢ =

8. 4¢ + 5¢ =

9. 3¢ + 3¢ =

10. 4¢ + 7¢ =

11.

6 rabbits	8 blue jays	3 fawns	3 bracelets	6 flowers
+ 2 rabbits	+ 4 blue jays	+ 9 fawns	+ 7 bracelets	+ 1 flower

12. 1 + 4

13. 2 + 5

14. 4 + 4

15. 3 + 9

16. 1 + 9

Answers

1. 5 pieces

Tip: Be sure to require fully worded answers.

2. 8 books

3. 8 pieces
4. 8 pounds
5. 9 eggs
6. 10¢
7. 10¢
8. 9¢
9. 6¢
10. 11¢
11.

6 rabbits	8 blue jays	3 fawns	3 bracelets	6 flowers
+ 2 rabbits	+ 4 blue jays	+ 9 fawns	+ 7 bracelets	+ 1 flower
8 rabbits	12 blue jays	12 fawns	10 bracelets	7 flowers

12. 5
13. 7
14. 8
15. 12
16. 10

Review 4 and Previous Tables: Subtraction

1. If Mark has 5¢ and he gives away 2¢, how much does he have left?

2. Lyla has 7 roses and gives away 4. How many roses are left?

3. Fiona saw 9 birds at the feeder, then 3 flew away. How many birds remain at the feeder?

4. I have 8 straws and pass out 4. How many straws do I have left?

5. There are 6 ducks swimming on the pond and 2 fly away. How many are left?

6. 4¢ – 2¢ =

7. 6¢ – 3¢ =

8. 8¢ – 3¢ =

9. 9¢ – 2¢ =

10. 6¢ – 4¢ =

11.

11 birds	5 deer	12 caterpillars	7 books	9 cookies
– 2 birds	– 3 deer	– 4 caterpillars	– 2 books	– 4 cookies

12. 5 – 1

13. 10 – 3

14. 4 – 4

15. 3 – 2

16. 11 – 4

Answers

1. 3¢

Tip: Be sure to require fully worded answers.

2. 3 roses
3. 6 birds
4. 4 straws
5. 4 ducks
6. 2¢
7. 3¢
8. 5¢
9. 7¢
10. 2¢
11.

11 birds	5 deer	12 caterpillars	7 books	9 cookies
– 2 birds	– 3 deer	– 4 caterpillars	– 2 books	– 4 cookies
9 birds	2 deer	8 caterpillars	5 books	5 cookies

12. 4

13. 7
14. 0
15. 1
16. 7

If your child needs more practice to be comfortable with this table, you can find additional review sums for 4 and previous tables on page 271. Also look at previous lessons for many more sums that you can use as is or by changing names, numbers, or objects.

Addition and Subtraction Table for 5

Tip: Keep lessons to 15 minutes of work on the newest concept and 5 minutes of review at the end, making each day's math time 20 minutes total.

Step 1: Assemble a table and speak it through.

Objects used: buttons

Tip: As the numbers get larger you may want your child to simply place one 5 to the left of the table, as illustrated below.

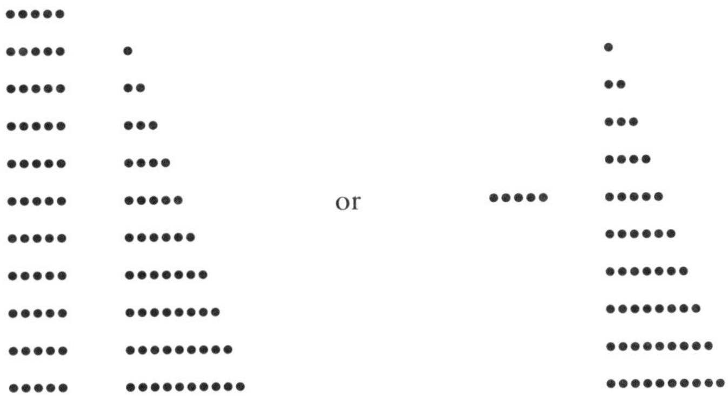

Tip: If you are working with an older child to solidify his knowledge of the

tables, and handwriting comes easily, the child may write the table in his math notebook rather than use objects to build it.

Step 2: Pose simple addition sums using the objects in the table.

Tip: After the initial day of building the new table, if you are not able to save it, then you may quickly build the table yourself before the lesson begins.

1. 5 buttons + 1 button =
2. 5 buttons + 5 buttons =
3. 5 buttons + 8 buttons =
4. 5 buttons + 4 buttons =
5. 5 buttons + 10 buttons =
6. 5 buttons + 3 buttons =
7. 5 buttons + 7 buttons =
8. 5 buttons + 9 buttons =
9. 5 buttons + 0 buttons =
10. 5 buttons + 2 buttons =
11. 5 buttons + 6 buttons =

Tip: If your child has advanced past the need to build a table, by all means put the objects away and continue the oral questions without the aid of the manipulatives. Simply get them back out if need be.

Answers

1. 6 buttons

Tip: Be sure to require fully worded answers: 5 + 1 = 6.

2. 10 buttons
3. 13 buttons
4. 9 buttons
5. 15 buttons
6. 8 buttons
7. 12 buttons
8. 14 buttons
9. 5 buttons
10. 7 buttons
11. 11 buttons

Step 3: Progress to addition sums with imaginary objects.

1. 5 dogs + 3 dogs = how many dogs?
2. 5 flowers + 4 flowers = how many flowers?
3. 5 rabbits + 6 rabbits = how many rabbits?
4. 5 books + 9 books = how many books?
5. 5 sponges + 1 sponge = how many sponges?
6. 5 pears + 8 pears = how many pears?
7. 5 teddy bears + 5 teddy bears = how many teddy bears?
8. 5 buckets + 2 buckets = how many buckets?
9. 5 gum balls + 10 gum balls = how many gum balls?
10. 5 ponies + 7 ponies = how many ponies?

Tip: Make sure answers are coming readily in the current step before you move to the next step.

Answers

1. 8 dogs

Tip: Be sure to require fully worded answers.

2. 9 flowers
3. 11 rabbits
4. 14 books
5. 6 sponges
6. 13 pears
7. 10 teddy bears
8. 7 buckets
9. 15 gum balls
10. 12 ponies

Step 4: Work with addition sums using pure number.

1. 5 + 5 =

2. 5 + 1 =

3. 5 + 8 =

4. 5 + 9 =

5. 5 + 2 =

6. 5 + 7 =

7. 5 + 4 =

8. 5 + 0 =

9. 5 + 10 =

10. 5 + 3 =

Answers

1. 10

Tip: Be sure to require fully worded answers.

2. 6
3. 13
4. 14
5. 7
6. 12
7. 9
8. 5
9. 15
10. 8

Step 5: Incorporate the subtraction table as addition continues.

1. 5 buttons added to 3 buttons equals how many buttons?

2. You have 5 buttons but you want 9 buttons. How many more must you get?

3. 5 buttons added to 5 buttons make how many buttons?

4. You have 5 buttons but want 7 buttons. How many more must you get?

5. 5 buttons added to 4 buttons make how many buttons?

6. You have 5 buttons and you want 10 buttons. How many more must you get?

7. 5 buttons added to 2 buttons equals how many buttons?

8. You have 5 buttons and want 12 buttons. How many more do you need?

9. 5 buttons added to 9 buttons make how many buttons?

10. You have 5 buttons but want 11 buttons. How many more buttons do you need?

11. 5 buttons added to 1 button make how many buttons?

12. 6 buttons minus 5 buttons equals how many buttons?

13. 5 buttons added to 10 buttons equals how many buttons?

14. You have 12 buttons and take 5 away. How many buttons remain?

15. 5 buttons added to 6 buttons make how many buttons?

16. You have 5 buttons but want 13 buttons. How many must you get?

17. 5 buttons added to 7 buttons equals how many buttons?

18. 8 buttons take away 5 buttons leaves how many buttons?

19. 5 buttons and 8 buttons together make how many buttons?

20. 14 buttons take away 5 buttons leaves you with how many buttons?

Answers

1. 8 buttons

Tip: Be sure to require fully worded answers.

2. 4 buttons
3. 10 buttons
4. 2 buttons
5. 9 buttons
6. 5 buttons
7. 7 buttons
8. 7 buttons
9. 14 buttons

10. 6 buttons
11. 6 buttons
12. 1 button
13. 15 buttons
14. 7 buttons
15. 11 buttons
16. 8 buttons
17. 12 buttons
18. 3 buttons
19. 13 buttons
20. 9 buttons

Step 6: Give mixed sums with imaginary objects.

Tip: On days when the work is going exceedingly well, you may have your child write on his slate or in his math notebook an addition or subtraction equation he has solved, using the proper signs.

1. 5 jacks from 9 jacks?
2. 5 bubbles from 7 bubbles?
3. 5 papers and 5 papers?
4. 5 peaches and 3 peaches?
5. 5 bunnies from 10 bunnies?
6. 5 dogs from 12 dogs?
7. 5 baskets and 1 basket?
8. 5 shoes from 6 shoes?
9. 5 books and 4 books?
10. 5 rabbits from 14 rabbits?

Answers

1. 4 jacks

Tip: Be sure to require fully worded answers.

2. 2 bubbles
3. 10 papers
4. 8 peaches
5. 5 bunnies
6. 7 dogs
7. 6 baskets
8. 1 shoe
9. 9 books
10. 9 rabbits

Step 7: Present mixed sums using pure number.

Tip: Be sure to put away the manipulatives when they are no longer needed for this table.

How much is left if you take

1. 5 from 5?
2. 5 from 7?
3. 5 from 10?
4. 5 from 9?
5. 5 from 12?

How much have you if you add

6. 5 and 1?
7. 5 and 7?
8. 5 and 5?
9. 5 and 8?
10. 5 and 4?

Answers

Tip: Be sure to require fully worded answers.

1. 0
2. 2

6. 6
7. 12

3. 5
4. 4
5. 7
8. 10
9. 13
10. 9

Review 5 with Previous Tables: Addition

1. Mom bought 4 notebooks for Harry and 3 notebooks for Jan. How many notebooks did she buy in all?

2. George has 5 balloons and was given 3 more. How many does he have?

3. Darius was given 3 pancakes for breakfast. There were 9 more still on the platter. How many pancakes were there in all?

4. Mary earned 4 pennies watching her baby brother and 2 pennies for cleaning the table. How many pennies did she earn all together?

5. Sarah used 3 crayons on her first picture and 7 crayons on her second. How many crayons did she use in all?

6. 3¢ + 4¢ =

7. 4¢ + 6¢ =

8. 2¢ + 3¢ =

9. 4¢ + 1¢ =

10. 5¢ + 5¢ =

11.

5 ducks	2 dogs	3 rabbits	4 flowers	5 chocolate chips
+ 4 ducks	+ 7 dogs	+ 6 rabbits	+ 9 flowers	+ 8 chocolate chips

12. 5 + 6

13. 4 + 10

14. 2 + 9

15. 4 + 5

16. 3 + 8

Answers

1. 7 notebooks

Tip: Be sure to require fully worded answers.

2. 8 balloons
3. 12 pancakes
4. 6 pennies
5. 10 crayons
6. 7¢
7. 10¢
8. 5¢
9. 5¢
10. 10¢
11.

5 ducks	2 dogs	3 rabbits	4 flowers	5 chocolate chips
+ 4 ducks	+ 7 dogs	+ 6 rabbits	+ 9 flowers	+ 8 chocolate chips
9 ducks	9 dogs	9 rabbits	13 flowers	13 chocolate chips

12. 11
13. 14
14. 11
15. 9
16. 11

Review 5 with Previous Tables: Subtraction

1. Daniel has 10 balloons and gives away 5 of them. How many balloons does he have?

2. Mary collected 12 seashells at the beach and gave 4 of them to her siblings. How many seashells does she have left?

3. Rachel has 8 colored pencils and shares 4 with her sister. How many does Rachel still have?

4. David's team scored 5 goals. David scored 2 of them. How many did his teammates score?

5. Sarah drew 6 pictures and gave 5 of them away. How many pictures does she have left?

6. 11¢ – 5¢ =

7. 7¢ – 4¢ =

8. 9¢ – 1¢ =

9. 4¢ – 3¢ =

10. 12¢ – 4¢ =

11.

9 pot holders	2 rulers	7 glasses	3 books	4 ribbons
– 4 pot holders	– 1 ruler	– 2 glasses	– 2 books	– 2 ribbons

12. 8 – 3

13. 5 – 4

14. 9 – 4

15. 6 – 3

16. 7 – 2

Answers

1. 5 balloons

Tip: Be sure to require fully worded answers.

2. 8 shells
3. 4 colored pencils
4. 3 goals
5. 1 picture
6. 6¢
7. 3¢
8. 8¢
9. 1¢
10. 8¢
11.

9 pot holders	2 rulers	7 glasses	3 books	4 ribbons
– 4 pot holders	– 1 ruler	– 2 glasses	– 2 books	– 2 ribbons
5 pot holders	1 ruler	5 glasses	1 book	2 ribbons

12. 5
13. 1
14. 5
15. 3
16. 5

If your child needs more practice to be comfortable with this table, you can find additional review sums for 5 and previous tables on page 273. Also look at previous lessons for many more sums that you can use as is or by changing names, numbers, or objects.

Addition and Subtraction Table for 6

Tip: As a help, pull a Number Sentence Card out before the lesson begins so your student's work can continue uninterrupted if you get called away for a moment.

Step 1: Assemble a table and speak it through.

Objects used: buttons

Tip: As the numbers get larger you may want your child to simply place one 6 to the left of the table, as illustrated below.

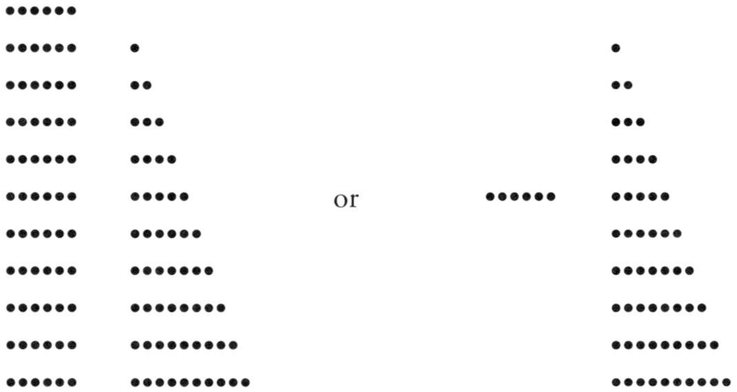

Tip: If you are working with an older child to solidify his knowledge of the tables, and handwriting comes easily, the child may write the table in his math notebook rather than use objects to build it.

Step 2: Pose simple addition sums using the objects in the table.

Tip: After the initial day of building the new table, if you are not able to save it, then you may quickly build the table yourself before the lesson begins.

1. 6 buttons + 6 buttons =
2. 6 buttons + 4 buttons =
3. 6 buttons + 5 buttons =
4. 6 buttons + 9 buttons =
5. 6 buttons + 3 buttons =
6. 6 buttons + 2 buttons =

7. 6 buttons + 10 buttons =

8. 6 buttons + 8 buttons =

9. 6 buttons + 1 button =

10. 6 buttons + 7 buttons =

11. 6 buttons + 0 buttons =

Tip: If your child has advanced past the need to build a table, put the objects away and continue the oral questions without the aid of the manipulatives. Simply get them back out if needed.

Answers

1. 12 buttons

   Tip: Be sure to require fully worded answers: 6 + 6 = 12.

2. 10 buttons
3. 11 buttons
4. 15 buttons
5. 9 buttons
6. 8 buttons
7. 16 buttons
8. 14 buttons
9. 7 buttons
10. 13 buttons
11. 6 buttons

Step 3: Progress to addition sums with imaginary objects.

1. 6 shirts + 5 shirts = how many shirts?

2. 6 ants + 7 ants = how many ants?

3. 6 blocks + 1 block = how many blocks?

4. 6 ducks + 8 ducks = how many ducks?

5. 6 pancakes + 2 pancakes = how many pancakes?

6. 6 foxes + 3 foxes = how many foxes?

7. 6 erasers + 10 erasers = how many erasers?

8. 6 lambs + 6 lambs = how many lambs?

9. 6 pencils + 4 pencils = how many pencils?

10. 6 acorns + 9 acorns = how many acorns?

Tip: Make sure answers are coming readily in the current step before you move to the next step.

Answers

1. 11 shirts

   Tip: Be sure to require fully worded answers.

2. 13 ants
3. 7 blocks
4. 14 ducks
5. 8 pancakes
6. 9 foxes
7. 16 erasers
8. 12 lambs
9. 10 pencils
10. 15 acorns

Step 4: Work with addition sums using pure number.

1. 6 + 5 =

2. 6 + 8 =

3. 6 + 4 =

4. 6 + 3 =

5. 6 + 1 =

6. 6 + 9 =

7. 6 + 7 =

8. 6 + 10 =

9. 6 + 2 =

10. 6 + 6 =

Answers

1. 11

Tip: Be sure to require fully worded answers.

2. 14
3. 10
4. 9
5. 7
6. 15
7. 13
8. 16
9. 8
10. 12

Step 5: Incorporate the subtraction table as addition continues.

1. 6 buttons added to 10 buttons make how many buttons?

2. You have 6 buttons but you want 14 buttons. How many more must you get?

3. 6 buttons added to 3 buttons make how many buttons?

4. You have 6 buttons but want 12 buttons. How many more must you get?

5. 6 buttons added to 5 buttons make how many buttons?

6. You have 6 buttons and you want 9 buttons. How many more must you get?

7. 6 buttons plus 1 button equals how many buttons?

8. If you have 6 buttons and want 11 buttons, how many more must you get?

9. 6 buttons added to 8 buttons make how many buttons?

10. You have 6 buttons but want 13 buttons. How many more buttons do you need?

11. 6 buttons added to 2 buttons make how many buttons?

12. 6 buttons take away 6 buttons leaves how many buttons?

13. 6 buttons plus 9 buttons equals how many buttons?

14. If you have 11 buttons and take 6 away, how many buttons do you have?

15. 6 buttons added to 7 buttons make how many buttons?

16. If you have 6 buttons but want 10 buttons, how many must you get?

17. 6 buttons added to 6 buttons make how many buttons?

18. 8 buttons take away 6 buttons leaves how many buttons?

19. 6 buttons and 4 buttons together make how many buttons?

20. 15 buttons minus 6 buttons leaves you with how many buttons?

Answers

1. 16 buttons

Tip: Be sure to require fully worded answers.

2. 8 buttons
3. 9 buttons
4. 6 buttons
5. 11 buttons
6. 3 buttons
7. 7 buttons
8. 5 buttons
9. 14 buttons
10. 7 buttons
11. 8 buttons
12. 0 buttons
13. 15 buttons
14. 5 buttons
15. 13 buttons
16. 4 buttons
17. 12 buttons
18. 2 buttons
19. 10 buttons
20. 9 buttons

Step 6: Give mixed sums with imaginary objects.

1. 6 ferns and 3 ferns?

2. 6 balloons from 9 balloons?

3. 6 feathers from 12 feathers?

4. 6 marbles and 7 marbles?

5. 6 straws and 10 straws?

6. 6 bunnies from 10 bunnies?

7. 6 pebbles and 8 pebbles?

8. 6 forks from 11 forks?

9. 6 toy cars and 1 toy car?

10. 8 ribbons take away 6 ribbons? *After solving, your student may write 8 – 6 = 2 in his math notebook.*

Answers

1. 9 ferns

Tip: Be sure to require fully worded answers.

2. 3 balloons
3. 6 feathers
4. 13 marbles
5. 16 straws
6. 4 bunnies
7. 14 pebbles
8. 5 forks
9. 7 cars
10. 2 ribbons

Step 7: Present mixed sums using pure number.

Tip: Be sure to put away the manipulatives when they are no longer needed for this table.

How much is left if you take	How much have you if you add
1. 6 from 12?	6. 6 and 4?
2. 6 from 9?	7. 6 and 7?
3. 6 from 7?	8. 6 and 10?
4. 6 from 11?	9. 6 and 3?
5. 6 from 6?	10. 6 and 9?

Answers

Tip: Be sure to require fully worded answers.

1. 6	6. 10
2. 3	7. 13
3. 1	8. 16
4. 5	9. 9
5. 0	10. 15

Review 6 and Previous Tables: Addition

1. Elia has 3 plums and Luke has 3 plums. How many plums have they together?

2. In the box are 6 brown puppies and 3 white puppies. How many puppies in all?

3. Elinor collected 5 eggs on Monday and 8 eggs on Tuesday. How many eggs did she collect all together?

4. Mrs. Schultz gathered 4 roses and 7 lilies. How many flowers has she gathered in all?

5. There are 6 hens in the coop and 7 hens outside. How many hens in all?

6. 4¢ + 3¢ =

7. 6¢ + 5¢ =

8. 2¢ + 8¢ =

9. 1¢ + 10¢ =

10. 5¢ + 7¢ =

11.

6 cows	3 pigs	4 hens	2 grapes	5 balls
+ 2 cows	+ 3 pigs	+ 2 hens	+ 9 grapes	+ 8 balls

12. 3 + 7

13. 5 + 4

14. 3 + 9

15. 4 + 6

16. 1 + 9

Answers

1. 6 plums

Tip: Be sure to require fully worded answers.

2. 9 puppies
3. 13 eggs
4. 11 flowers
5. 13 hens
6. 7¢
7. 11¢
8. 10¢
9. 11¢
10. 12¢

11.

6 cows	3 pigs	4 hens	2 grapes	5 balls
+ 2 cows	+ 3 pigs	+ 2 hens	+ 9 grapes	+ 8 balls
8 cows	6 pigs	6 hens	11 grapes	13 balls

12. 10
13. 9
14. 12
15. 10
16. 10

Review 6 and Previous Tables: Subtraction

1. Robbie makes 6 corner bookmarks and gives away 4. How many does he have left?

2. There are 10 magnets on the refrigerator and baby brother pulls 1 of them off. How many remain?

3. On our nature walk yesterday, we saw 8 ferns; today we saw 6. How many ferns all together?

4. We gathered 14 paintbrushes for our project. We used only 6. How many paintbrushes do we have left?

5. Aviyah is working on her embroidery project. She has 10 stitches to accomplish and 4 are complete. How many more must she finish?

6. 4¢ – 1¢ =

7. 10¢ – 6¢ =

8. 7¢ – 4¢ =

9. 6¢ – 3¢ =

10. 6¢ – 5¢ =

11.

6 papers	11 arrowheads	8 fish	13 ornaments	5 flowers
– 6 papers	– 4 arrowheads	– 6 fish	– 3 ornaments	– 3 flowers

12. 9 – 3

13. 5 – 4

14. 7 – 2

15. 11 – 5

16. 12 – 3

Answers

1. 2 bookmarks

Tip: Be sure to require fully worded answers.

2. 9 magnets
3. 14 ferns
4. 8 paintbrushes
5. 6 stitches
6. 3¢
7. 4¢
8. 3¢
9. 3¢
10. 1¢
11.

6 papers	11 arrowheads	8 fish	13 ornaments	5 flowers
– 6 papers	– 4 arrowheads	– 6 fish	– 3 ornaments	– 3 flowers
0 papers	7 arrowhead	2 fish	10 ornaments	2 flowers

12. 6
13. 1
14. 5
15. 6
16. 9

If your child needs more practice to be comfortable with this table, you can find additional review sums for 6 and previous tables on page 275. Also look at previous lessons for many more sums that you can use as is or by changing names, numbers, or objects.

Addition and Subtraction Table for 7

Tip: Keep lessons to 15 minutes of work on the newest concept and 5 minutes of review at the end, making each day's math time 20 minutes total.

Step 1: Assemble a table and speak it through.

Objects used: pennies

Tip: As the numbers get larger you may want your child to simply place one 7 to the left of the table, as illustrated below.

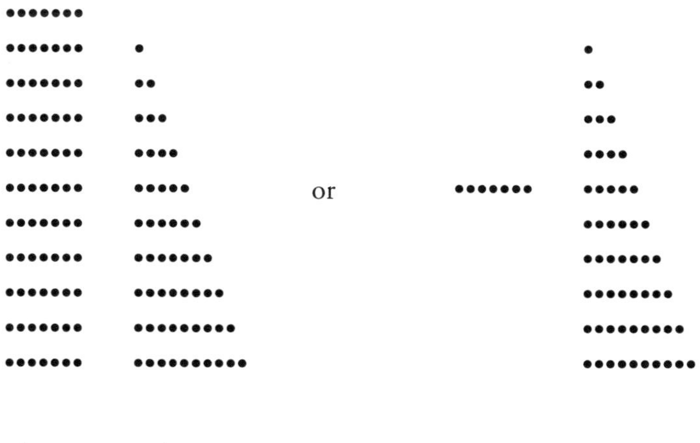

Tip: If you are working with an older child to solidify his knowledge of the tables, and handwriting comes easily, the child may write the table in his math notebook rather than use objects to build it.

Step 2: Pose simple addition sums using the objects in the table.

Tip: After the initial day of building the new table, if you are not able to save it, then you may quickly build the table yourself before the lesson begins.

1. 7 pennies + 1 penny =
2. 7 pennies + 3 pennies =
3. 7 pennies + 10 pennies =
4. 7 pennies + 6 pennies =
5. 7 pennies + 8 pennies =
6. 7 pennies + 4 pennies =
7. 7 pennies + 2 pennies =
8. 7 pennies + 7 pennies =
9. 7 pennies + 0 pennies =
10. 7 pennies + 5 pennies =
11. 7 pennies + 9 pennies =

Answers

1. 8 pennies

Tip: Be sure to require fully worded answers: 7 + 1 = 8.

2. 10 pennies
3. 17 pennies
4. 13 pennies
5. 15 pennies
6. 11 pennies

7. 9 pennies
8. 14 pennies
9. 7 pennies
10. 12 pennies
11. 16 pennies

Step 3: Progress to addition sums with imaginary objects.

1. 7 rabbits + 2 rabbits = how many rabbits?
2. 7 ducks + 6 ducks = how many ducks?
3. 7 pencils + 9 pencils = how many pencils?
4. 7 boats + 4 boats = how many boats?
5. 7 stamps + 1 stamp = how many stamps?
6. 7 stickers + 10 stickers = how many stickers?
7. 7 books + 8 books = how many books?
8. 7 balls + 3 balls = how many balls?
9. 7 cucumbers + 5 cucumbers = how many cucumbers?
10. 7 clouds + 7 clouds = how many clouds?

Tip: Make sure answers are coming readily in the current step before you move to the next step.

Answers

1. 9 rabbits

Tip: Be sure to require fully worded answers.

2. 13 ducks

3. 16 pencils
4. 11 boats
5. 8 stamps
6. 17 stickers
7. 15 books
8. 10 balls
9. 12 cucumbers
10. 14 clouds

Step 4: Work with addition sums using pure number.

Tip: Remember, manipulatives are put away when no longer necessary.

1. 7 + 7 =

2. 7 + 3 =

3. 7 + 4 =

4. 7 + 9 =

5. 7 + 5 =

6. 7 + 2 =

7. 7 + 0 =

8. 7 + 10 =

9. 7 + 6 =

10. 7 + 8 =

Answers

1. 14

Tip: Be sure to require fully worded answers.

2. 10
3. 11
4. 16
5. 12
6. 9
7. 7
8. 17
9. 13
10. 15

Step 5: Incorporate the subtraction table as addition continues.

1. 7 pennies added to 5 pennies make how many pennies?
2. 7 pennies take away 7 pennies leaves how many pennies?
3. 7 pennies added to 8 pennies equals how many pennies?
4. If you have 9 pennies and take 7 away, how many pennies do you have?
5. 7 pennies added to 9 pennies make how many pennies?
6. If you have 7 pennies but want 15 pennies, how many must you get?
7. 7 pennies added to 0 pennies make how many pennies?
8. 7 pennies added to 1 penny make how many pennies?
9. You have 7 pennies but want 13 pennies. How many more must you get?
10. 7 pennies added to 4 pennies make how many pennies?

11. You have 7 pennies and you want 17 pennies. How many more must you get?

12. 7 pennies added to 3 pennies make how many pennies?

13. If you have 7 pennies and want 8 pennies, how many more must you get?

14. 7 pennies added to 6 pennies make how many pennies?

15. You have 7 pennies but want 14 pennies. How many more pennies do you need?

16. 11 pennies take away 7 pennies leaves how many pennies?

17. 7 pennies and 2 pennies together make how many pennies?

18. 12 pennies take away 7 pennies leaves you with how many pennies?

19. 7 pennies added to 7 pennies make how many pennies?

20. You have 7 pennies but you want 16 pennies. How many more must you get?

Answers

1. 12 pennies

Tip: Be sure to require fully worded answers.

2. 0 pennies
3. 15 pennies
4. 2 pennies
5. 16 pennies
6. 8 pennies
7. 7 pennies
8. 8 pennies
9. 6 pennies
10. 11 pennies

11. 10 pennies
12. 10 pennies
13. 1 penny
14. 13 pennies
15. 7 pennies
16. 4 pennies
17. 9 pennies
18. 5 pennies
19. 14 pennies
20. 9 pennies

Step 6: Give mixed sums with imaginary objects.

Tip: On days when the work is going exceedingly well, you may have your child write on his slate or in his math notebook an addition or subtraction equation he has solved, using the proper signs.

1. 7 bunnies from 14 bunnies?

2. 7 frogs from 9 frogs?

3. 7 books and 5 books?

4. 7 eggs and 3 eggs?

5. 7 straws from 10 straws?

6. 7 papers from 12 papers?

7. 7 dogs plus 4 dogs?

8. 7 napkins from 15 napkins?

9. 7 flowers and 8 flowers?

10. 7 cicadas from 8 cicadas?

Answers

1. 7 bunnies

Tip: Be sure to require fully worded answers.

2. 2 frogs
3. 12 books
4. 10 eggs
5. 3 straws
6. 5 papers
7. 11 dogs
8. 8 napkins
9. 15 flowers
10. 1 cicada

Step 7: Present mixed sums using pure number.

Tip: Be sure to put away the manipulatives when they are no longer needed for this table.

How much is left if you take

1. 7 from 12?
2. 7 from 10?
3. 7 from 8?
4. 7 from 16?
5. 7 from 15?

How much have you if you add

6. 7 and 5?
7. 7 and 8?
8. 7 and 4?
9. 7 and 10?
10. 7 and 1?

Answers

Tip: Be sure to require fully worded answers.

1. 5
2. 3

6. 12
7. 15

3. 1
4. 9
5. 8
8. 11
9. 17
10. 8

Review 7 and Previous Tables: Addition

1. There are 6 leaves on the ground. The next day there are 3 more. How many leaves are on the ground?

2. The Barnes family has 1 tree planted in their yard. They decide to plant 4 more. How many trees will they have in their yard?

3. There are 4 chocolate chip cookies on the plate. 7 more come out of the oven. How many cookies are there in all?

4. 2 yellow trucks go by and later 5 red cars go by. How many vehicles went by all together?

5. There are 7 rabbits nibbling grass in the yard. 3 more join them. How many rabbits are in the yard?

6. 3¢ + 4¢ =

7. 7¢ + 6¢ =

8. 2¢ + 4¢ =

9. 5¢ + 4¢ =

10. 6¢ + 6¢ =

11.

4 toads	3 unicorns	5 falcons	6 hawks	7 pinecones
+ 6 toads	+ 3 unicorns	+ 10 falcons	+ 5 hawks	+ 9 pinecones

12. 7 + 7

13. 4 + 8

14. 6 + 5

15. 3 + 9

16. 2 + 6

Answers

1. 9 leaves

Tip: Be sure to require fully worded answers.

2. 5 trees
3. 11 cookies
4. 7 vehicles
5. 10 rabbits
6. 7¢
7. 13¢
8. 6¢
9. 9¢
10. 12¢
11.

4 toads	3 unicorns	5 falcons	6 hawks	7 pinecones
+ 6 toads	+ 3 unicorns	+ 10 falcons	+ 5 hawks	+ 9 pinecones
10 toads	6 unicorns	15 falcons	11 hawks	16 pinecones

12. 14
13. 12
14. 11
15. 12
16. 8

Review 7 and Previous Tables: Subtraction

1. If I have 7 pins and give away 4, how many do I have left?

2. We need 6 stamps for our package, but only have 4. How many more stamps do we need?

3. There are 7 puppets onstage and 2 leave. How many are left?

4. Mr. Beal caught 11 fish. He sold 7 of them. How many does he have left?

5. The team needs 10 hats for the ball game but only has 7. How many more are needed?

6. 15¢ – 6¢ =

7. 8¢ – 5¢ =

8. 11¢ – 4¢ =

9. 17¢ – 7¢ =

10. 12¢ – 3¢ =

11.

17 magnets	6 drums	9 cookies	8 mice	9 pigs
– 7 magnets	– 5 drums	– 3 cookies	– 4 mice	– 7 pigs

12. 15 – 7

13. 13 – 6

14. 14 – 4

15. 11 – 3

16. 14 – 7

Answers

1. 3 pins

Tip: Be sure to require fully worded answers.

2. 2 stamps
3. 5 puppets

4. 4 fish
5. 3 hats
6. 9¢
7. 3¢
8. 7¢
9. 10¢
10. 9¢
11.

17 magnets	6 drums	9 cookies	8 mice	9 pigs
– 7 magnets	– 5 drums	– 3 cookies	– 4 mice	– 7 pigs
10 magnets	1 drum	6 cookies	4 mice	2 pigs

12. 8
13. 7
14. 10
15. 8
16. 7

If your child needs more practice to be comfortable with this table, you can find additional review sums for 7 and previous tables on page 277. Also look at previous lessons for many more sums that you can use as is or by changing names, numbers, or objects.

Addition and Subtraction Table for 8

Tip: As a help, pull a Number Sentence Card out before the lesson begins so your student's work can continue uninterrupted if you get called away for a moment.

Step 1: Assemble a table and speak it through.

Objects used: pennies

Tip: As the numbers get larger you may want your child to simply place one 8 to the left of the table, as illustrated below.

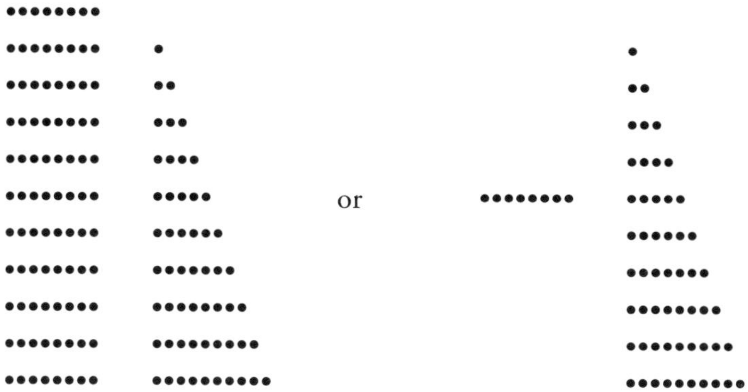

Tip: If you are working with an older child to solidify his knowledge of the tables, and handwriting comes easily, the child may write the table in his math notebook rather than use objects to build it.

Step 2: Pose simple addition sums using the objects in the table.

Tip: After the initial day of building the new table, if you are not able to save it, then you may quickly build the table yourself before the lesson begins.

1. 8 pennies + 4 pennies =
2. 8 pennies + 9 pennies =
3. 8 pennies + 5 pennies =
4. 8 pennies + 3 pennies =
5. 8 pennies + 1 penny =
6. 8 pennies + 7 pennies =
7. 8 pennies + 0 pennies =

8. 8 pennies + 10 pennies =

9. 8 pennies + 2 pennies =

10. 8 pennies + 8 pennies =

11. 8 pennies + 6 pennies =

Answers

1. 12 pennies

Tip: Be sure to require fully worded answers: 8 + 4 = 12.

2. 17 pennies
3. 13 pennies
4. 11 pennies
5. 9 pennies
6. 15 pennies
7. 8 pennies
8. 18 pennies
9. 10 pennies
10. 16 pennies
11. 14 pennies

Step 3: Progress to addition sums with imaginary objects.

1. 8 bunnies + 4 bunnies = how many bunnies?

2. 8 pebbles + 10 pebbles = how many pebbles?

3. 8 chickens + 8 chickens = how many chickens?

4. 8 baskets + 6 baskets = how many baskets?

5. 8 grapes + 9 grapes = how many grapes?

6. 8 chairs + 2 chairs = how many chairs?

7. 8 lamps + 1 lamp = how many lamps?

8. 8 bubbles + 3 bubbles = how many bubbles?

9. 8 popsicles + 5 popsicles = how many popsicles?

10. 8 bottles + 7 bottles = how many bottles?

Tip: Make sure answers are coming readily in the current step before you move to the next step.

Answers

1. 12 bunnies

Tip: Be sure to require fully worded answers.

2. 18 pebbles
3. 16 chickens
4. 14 baskets
5. 17 grapes
6. 10 chairs
7. 9 lamps
8. 11 bubbles
9. 13 popsicles
10. 15 bottles

Step 4: Work with addition sums using pure number.

1. 8 + 5 =

2. 8 + 9 =

3. 8 + 4 =

4. 8 + 2 =

5. 8 + 7 =

6. 8 + 1 =

7. 8 + 8 =

8. 8 + 10 =

9. 8 + 6 =

10. 8 + 3 =

Answers

1. 13

Tip: Be sure to require fully worded answers.

2. 17
3. 12
4. 10
5. 15
6. 9
7. 16
8. 18
9. 14
10. 11

Step 5: Incorporate the subtraction table as addition continues.

1. 8 pennies added to 5 pennies make how many pennies?

2. You have 8 pennies but want 16 pennies. How many more pennies do you need?

3. 8 pennies added to 7 pennies make how many pennies?

4. 11 pennies take away 8 pennies leaves how many pennies?

5. 8 pennies added to 4 pennies equals how many pennies?

6. If you have 17 pennies and take 8 away, how many pennies do you have?

7. You have 8 pennies but want 13 pennies. How many more must you get?

8. 8 pennies added to 3 pennies make how many pennies?

9. You have 8 pennies and you want 14 pennies. How many more must you get?

10. 8 pennies plus 1 penny equals how many pennies?

11. You have 8 pennies and want 8 pennies. How many more must you get?

12. 8 pennies plus 10 pennies equals make how many pennies?

13. 8 pennies added to 9 pennies make how many pennies?

14. 8 pennies and 8 pennies together make how many pennies?

15. 12 pennies minus 8 pennies equals how many pennies?

16. If you have 8 pennies but want 10 pennies, how many more do you need?

17. 8 pennies added to 6 pennies make how many pennies?

18. 9 pennies minus 8 pennies equals how many pennies?

19. 8 pennies added to 2 pennies make how many pennies?

20. 8 pennies taken from 15 pennies leaves how many pennies?

Answers

1. 13 pennies

Tip: Be sure to require fully worded answers.

2. 8 pennies
3. 15 pennies
4. 3 pennies
5. 12 pennies
6. 9 pennies
7. 5 pennies
8. 11 pennies
9. 6 pennies
10. 9 pennies
11. 0 pennies
12. 18 pennies
13. 17 pennies
14. 16 pennies
15. 4 pennies
16. 2 pennies
17. 14 pennies
18. 1 penny
19. 10 pennies
20. 7 pennies

Step 6: Give mixed sums with imaginary objects.

1. 8 popsicles and 4 popsicles?

2. 8 pinecones from 10 pinecones?

3. 8 waffles and 8 waffles?

4. 8 shells from 14 shells?

5. 8 birds and 9 birds?

6. 8 whistles from 13 whistles?

7. 8 pins and 3 pins?

8. 8 beads from 15 beads?

9. 8 fossils and 2 fossils? *After solving, he may write 8 + 2 = 10 on either his slate or in his math notebook.*

10. 8 crayons from 16 crayons? *After solving, he may write 16 – 8 = 8 or 8 + 8 = 16 on his slate or in his math notebook.*

Answers

1. 12 popsicles

Tip: Be sure to require fully worded answers.

2. 2 pinecones
3. 16 waffles
4. 6 shells
5. 17 birds
6. 5 whistles
7. 11 pins
8. 7 beads
9. 10 fossils
10. 8 crayons

Step 7: Present mixed sums using pure number.

Tip: Be sure to put away the manipulatives when they are no longer needed for this table.

How much is left if you take

1. 8 from 14?

2. 8 from 10?

How much have you if you add

6. 8 and 2?

7. 8 and 9?

3. 8 from 9?

4. 8 from 11?

5. 8 from 15?

8. 8 and 3?

9. 8 and 7?

10. 8 and 4?

Answers

Tip: Be sure to require fully worded answers.

1. 6
2. 2
3. 1
4. 3
5. 7
6. 10
7. 17
8. 11
9. 15
10. 12

Review 8 and Previous Tables: Addition

1. There are 4 squirrels gathering nuts for the winter. 4 more join them. How many squirrels are preparing for winter?

2. There are 8 pinecones on the ground and 5 more fall to the ground. How many pinecones are on the ground?

3. 6 goldfinches are eating at the feeder. 3 more goldfinches join them. How many total goldfinches are at the feeder?

4. Daniel read 4 books over the weekend and 5 more books throughout the week. How many books did Daniel read all together?

5. Jane gave 3 presents to her neighbors. Later she gave 1 more. How many gifts did she give in all?

6. 6¢ + 4¢ =

7. 5¢ + 3¢ =

8. 7¢ + 9¢ =

9. 6¢ + 2¢ =

10. 3¢ + 5¢ =

11.

8 guitars	4 blankets	7 moths	4 horses	7 bags
+ 2 guitars	+ 4 blankets	+ 2 moths	+ 3 horses	+ 5 bags

12. 8 + 4

13. 3 + 9

14. 7 + 6

15. 8 + 7

16. 2 + 5

Answers

1. 8 squirrels

Tip: Be sure to require fully worded answers.

2. 13 pinecones
3. 9 goldfinches
4. 9 books
5. 4 presents
6. 10¢
7. 8¢
8. 16¢
9. 8¢
10. 8¢
11.

8 guitars	4 blankets	7 moths	4 bags	7 bags
+ 2 guitars	+ 4 blankets	+ 2 moths	+ 3 horses	+ 5 bags
10 guitars	8 blankets	9 moths	7 horses	12 bags

12. 12

13. 12
14. 13
15. 15
16. 7

Review 8 and Previous Tables: Subtraction

1. George has 9 balloons and 4 pop. How many balloons does he have left?

2. Jack is putting together a 15-piece puzzle and has completed 6 pieces. How many does he have left?

3. Sarina is cleaning up her room and finds 4 hair clips. She owns a total of 6. How many is she still missing?

4. A farmer having 16 cows sold 8 of them. How many cows does he still own?

5. Jerry has 8¢ to buy a bone for his dog. The bone costs 11¢. How much more does he need?

6. 10¢ – 2¢ =

7. 15¢ – 7¢ =

8. 11¢ – 6¢ =

9. 14¢ – 5¢ =

10. 11¢ – 7¢ =

11.

8 whistles	12 pens	5 erasers	17 cherries	9 pelts
– 5 whistles	– 3 pens	– 1 eraser	– 8 cherries	– 1 pelt

12. 9 – 8

13. 16 – 7

14. 14 – 6

15. 11 – 4

16. 9 – 7

Answers

1. 5 balloons

Tip: Be sure to require fully worded answers.

2. 9 pieces
3. 2 hair clips
4. 8 cows
5. 3¢
6. 8¢
7. 8¢
8. 5¢
9. 9¢
10. 4¢
11.

8 whistles	12 pens	5 erasers	17 cherries	9 pelts
– 5 whistles	– 3 pens	– 1 eraser	– 8 cherries	– 1 pelt
3 whistles	9 pens	4 erasers	9 cherries	8 pelts

12. 1
13. 9
14. 8
15. 7
16. 2

If your child needs more practice to be comfortable with this table, you can find additional review sums for 8 and previous tables on page 279. Also look at previous lessons for many more sums that you can use as is or by changing names, numbers, or objects.

Addition and Subtraction Table for 9

Tip: Keep lessons to 15 minutes of work on the newest concept and 5 minutes of review at the end, making each day's math time 20 minutes total.

Step 1: Assemble a table and speak it through.

Objects used: beans

Tip: As the numbers get larger you may want your child to simply place one 9 to the left of the table, as illustrated below.

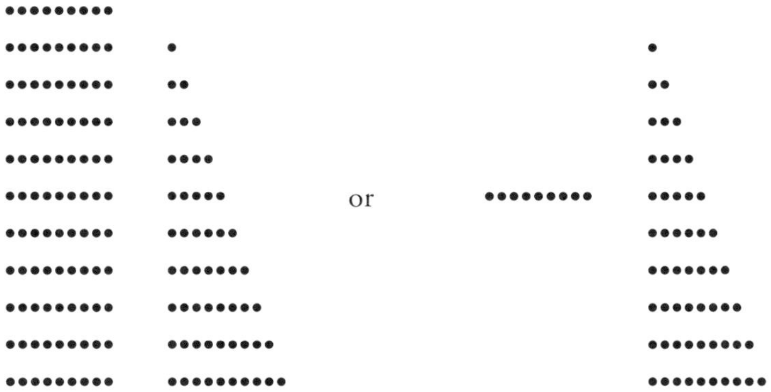

Tip: If you are working with an older child to solidify his knowledge of the tables, and handwriting comes easily, the child may write the table in his math notebook rather than use objects to build it.

Step 2: Pose simple addition sums using the objects in the table.

Tip: After the initial day of building the new table, if you are not able to save it, then you may quickly build the table yourself before the lesson begins.

1. 9 beans + 1 bean =

2. 9 beans + 5 beans =

3. 9 beans + 2 beans =

4. 9 beans + 0 beans =

5. 9 beans + 6 beans =

6. 9 beans + 3 beans =

7. 9 beans + 9 beans =

8. 9 beans + 8 beans =

9. 9 beans + 4 beans =

10. 9 beans + 7 beans =

11. 9 beans + 10 beans =

Answers

1. 10 beans

Tip: Be sure to require fully worded answers: 9 + 1 = 10.

2. 14 beans
3. 11 beans
4. 9 beans
5. 15 beans
6. 12 beans
7. 18 beans
8. 17 beans
9. 13 beans
10. 16 beans
11. 19 beans

Step 3: Progress to addition sums with imaginary objects.

1. 9 salamanders + 5 salamanders = how many salamanders?

2. 9 bears + 6 bears = how many bears?

3. 9 pencil + 7 pencils = how many pencils?

4. 9 toes + 1 toe = how many toes?

5. 9 ducks + 4 ducks = how many ducks?

6. 9 stamps + 9 stamps = how many stamps?

7. 9 jelly beans + 10 jelly beans = how many jelly beans?

8. 9 rocks + 8 rocks = how many rocks?

9. 9 socks + 3 socks = how many socks?

10. 9 birds + 2 birds = how many birds?

Tip: Make sure answers are coming readily in the current step before you move to the next step.

Answers

1. 14 salamanders

Tip: Be sure to require fully worded answers.

2. 15 bears
3. 16 pencils
4. 10 toes
5. 13 ducks
6. 18 stamps
7. 19 jelly beans
8. 17 rocks
9. 12 socks
10. 11 birds

Step 4: Work with addition sums using pure number.

1. 9 + 5 =
2. 9 + 2 =
3. 9 + 9 =
4. 9 + 4 =
5. 9 + 1 =
6. 9 + 7 =
7. 9 + 8 =
8. 9 + 10 =
9. 9 + 3 =
10. 9 + 6 =

Answers

1. 14

Tip: Be sure to require fully worded answers.

2. 11
3. 18
4. 13
5. 10
6. 16
7. 17
8. 19
9. 12
10. 15

Step 5: Incorporate the subtraction table as addition continues.

1. 9 beans added to 2 beans make how many beans?
2. 9 beans plus 3 beans equals how many beans?
3. You have 9 beans but you want 19 beans. How many more must you get?
4. 9 beans added to 9 beans make how many beans?
5. You have 9 beans and want 12 beans. How many more must you get?
6. 9 beans added to 8 beans make how many beans?
7. 13 beans take away 9 beans leaves how many beans?
8. 9 beans and 1 bean make how many beans?
9. 9 beans minus 9 beans equals how many beans?
10. 9 beans added to 4 beans equals how many beans?
11. You have 11 beans and take 9 away. How many beans have you now?
12. 9 beans added to 6 beans make how many beans?
13. You have 9 beans but want 14 beans. How many more must you get?
14. 9 beans plus 5 beans equals how many beans?
15. You have 9 beans and you want 16 beans. How many more must you get?
16. 9 beans and 10 beans make how many beans?

17. You have 9 beans but want 10 beans. How many must you get?

18. 9 beans added to 7 beans make how many beans?

19. 15 beans take away 9 beans leaves how many beans?

20. 18 beans minus 9 beans equals how many beans?

Answers

1. 11 beans

Tip: Be sure to require fully worded answers.

2. 12 beans
3. 10 beans
4. 18 beans
5. 3 beans
6. 17 beans
7. 4 beans
8. 10 beans
9. 0 beans
10. 13 beans
11. 2 beans
12. 15 beans
13. 5 beans
14. 14 beans
15. 7 beans
16. 19 beans
17. 1 bean
18. 16 beans
19. 6 beans
20. 9 beans

Step 6: Give mixed sums with imaginary objects.

Tip: On days when the work is going exceedingly well, you may have your child write on his slate or in his math notebook an addition or subtraction equation he has solved, using the proper signs.

1. 9 deer from 14 deer?
2. 9 dogs and 1 dog?
3. 9 marbles and 2 marbles?
4. 9 squares and 7 squares?
5. 9 strawberries from 10 strawberries?
6. 9 balloons and 8 balloons?
7. 9 puzzle pieces and 4 puzzle pieces?
8. 9 robins from 9 robins?
9. 9 foxes from 17 foxes?
10. 9 roses and 3 roses?

Answers

1. 5 deer

Tip: Be sure to require fully worded answers.

2. 10 dogs
3. 11 marbles
4. 16 squares
5. 1 strawberry
6. 17 balloons
7. 13 puzzle pieces
8. 0 robins
9. 8 foxes
10. 12 roses

Step 7: Present mixed sums using pure number.

Tip: Be sure to put away the manipulatives when they are no longer needed for this table.

How much is left if you take

1. 9 from 9?
2. 9 from 10?
3. 9 from 14?
4. 9 from 19?
5. 9 from 17?

How much have you if you add

6. 9 and 1?
7. 9 and 5?
8. 9 and 4?
9. 9 and 9?
10. 9 and 7?

Answers

Tip: Be sure to require fully worded answers.

1. 0
2. 1
3. 5
4. 10
5. 8
6. 10
7. 14
8. 13
9. 18
10. 16

Review 9 and Previous Tables: Addition

1. 5 bunnies hop across the field. 5 more join them. How many bunnies in all?

2. We have 6 chairs around our dinner table but we need 4 more for our guests. How many chairs do we need total?

3. Andrew is planning a canoe trip. He must drive 1 mile to the river,

canoe 4 miles, and drive 1 mile back. How many miles will he travel in all?

Tip: This idea is taken a bit experimentally. Every so often in this book, you will find an experimental sum that is a little more abstract in nature, just to see whether your child is ready to grasp it yet. Stay relaxed, and think of these experimental sums as "knocking on the door" to see if it is ready to open. Some children will get it; some won't right now. If your child doesn't comprehend an experimental idea, don't worry about it; simply move on to a different sum for now.

4. For nature study, Jack found 5 mushrooms and Ryan found 4 different mushrooms. How many mushrooms did the boys observe in all?

5. There are 8 goldfish swimming inside Samuel's aquarium. Samuel also has 7 other fish. How many fish does he have in the aquarium all together?

6. 9¢ + 5¢ =

7. 3¢ + 8¢ =

8. 6¢ + 6¢ =

9. 1¢ + 4¢ =

10. 6¢ + 9¢ =

11.

7 trees	5 ladders	9 signs	8 balls	9 swallows
+ 4 trees	+ 1 ladder	+ 2 signs	+ 6 balls	+ 6 swallows

12. 8 + 4

13. 3 + 9

14. 7 + 6

15. 8 + 8

16. 2 + 5

Answers

1. 10 bunnies

Tip: Be sure to require fully worded answers.

2. 10 chairs
3. 6 miles
4. 9 mushrooms
5. 15 fish
6. 14¢
7. 11¢
8. 12¢
9. 5¢
10. 15¢
11.

7 trees	5 ladders	9 signs	8 balls	9 swallows
+ 4 trees	+ 1 ladder	+ 2 signs	+ 6 balls	+ 6 swallows
11 trees	6 ladders	11 signs	14 balls	15 swallows

12. 12
13. 12
14. 13
15. 16
16. 7

Review 9 and Previous Tables: Subtraction

1. Jackie jumped roped 14 times her first try and only 7 jumps her second try. How many more jumps did Jackie do her first try than her second?

2. If John buys 13 gummy bears and eats 6, how many does he have left?

3. Of 12 fish in the aquarium, we can see only 7. How many fish are hidden?

4. We need 6 stamps for our letters, but only have 4. How many more do we need?

5. We made 10 cinnamon rolls for our birthday celebration and ate 6. How many do we have left?

6. 15¢ – 9¢ =

7. 12¢ – 9¢ =

8. 11¢ – 3¢ =

9. 16¢ – 7¢ =

10. 11¢ – 6¢ =

11.

18 rulers	10 bones	7 crayons	11 raisins	14 ferns
– 9 rulers	– 4 bones	– 6 crayons	– 4 raisins	– 9 ferns

12. 19 – 9

13. 11 – 2

14. 13 – 5

15. 17 – 7

16. 16 – 8

Answers

1. 7 jumps

Tip: Be sure to require fully worded answers.

2. 7 gummy bears

3. 5 fish
4. 2 stamps
5. 4 cinnamon rolls
6. 6¢
7. 3¢
8. 8¢
9. 9¢
10. 5¢
11.

18 rulers	10 bones	7 crayons	11 raisins	14 ferns
– 9 rulers	– 4 bones	– 6 crayons	– 4 raisins	– 9 ferns
9 rulers	6 bones	1 crayon	7 raisins	5 ferns

12. 10
13. 9
14. 8
15. 10
16. 8

If your child needs more practice to be comfortable with this table, you can find additional review sums for 9 and previous tables on page 281. Also look at previous lessons for many more sums that you can use as is or by changing names, numbers, or objects.

Addition and Subtraction Table for 10

Tip: As a help, pull a Number Sentence Card out before the lesson begins so your student's work can continue uninterrupted if you get called away for a moment.

Step 1: Assemble a table and speak it through.

Objects used: beans

Tip: As the numbers get larger you may want your child to simply place one 10 to the left of the table, as illustrated below.

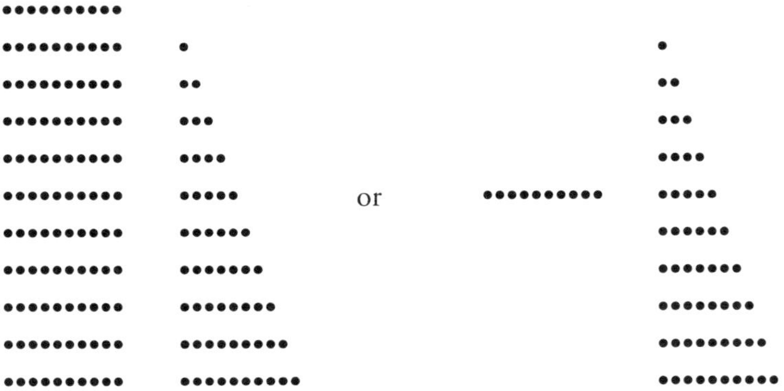

Tip: If you are working with an older child to solidify his knowledge of the tables, and handwriting comes easily, the child may write the table in his math notebook rather than use objects to build it.

Step 2: Pose simple addition sums using the objects in the table.

Tip: After the initial day of building the new table, if you are not able to save it, then you may quickly build the table yourself before the lesson begins.

1. 10 beans + 3 beans =

2. 10 beans + 6 beans =

3. 10 beans + 7 beans =

4. 10 beans + 4 beans =

5. 10 beans + 10 beans =

6. 10 beans + 5 beans =

7. 10 beans + 1 bean =

8. 10 beans + 9 beans =

9. 10 beans + 2 beans =

10. 10 beans + 8 beans =

11. 10 beans + 0 beans =

Answers

1. 13 beans

Tip: Be sure to require fully worded answers: 10 + 3 = 13.

2. 16 beans
3. 17 beans
4. 14 beans
5. 20 beans
6. 15 beans
7. 11 beans
8. 19 beans
9. 12 beans
10. 18 beans
11. 10 beans

Step 3: Progress to addition sums with imaginary objects.

1. 10 water bottles + 4 water bottles = how many water bottles?

2. 10 coolers + 2 coolers = how many coolers?

3. 10 dollars + 10 dollars = how many dollars?

4. 10 knives + 1 knife = how many knives?

5. 10 hairbands + 9 hairbands = how many hairbands?

6. 10 statues + 5 statues = how many statues?

7. 10 chairs + 7 chairs = how many chairs?

8. 10 balls + 8 balls = how many balls?

9. 10 fingers + 0 fingers = how many fingers?

10. 10 flowers + 3 flowers = how many flowers?

Tip: Make sure answers are coming readily in the current step before you move to the next step.

Answers

1. 14 water bottles

Tip: Be sure to require fully worded answers.

2. 12 coolers
3. 20 dollars
4. 11 knives
5. 19 hairbands
6. 15 statues
7. 17 chairs
8. 18 balls
9. 10 fingers
10. 13 flowers

Step 4: Work with addition sums using pure number.

1. 10 + 4 =

2. 10 + 6 =

3. 10 + 8 =

4. 10 + 9 =

5. 10 + 3 =

6. 10 + 2 =

7. 10 + 5 =

8. 10 + 10 =

9. 10 + 7 =

10. 10 + 1 =

Answers

1. 14

Tip: Be sure to require fully worded answers.

2. 16
3. 18
4. 19
5. 13
6. 12
7. 15
8. 20
9. 17
10. 11

Step 5: Incorporate the subtraction table as addition continues.

1. 10 beans added to 3 beans make how many beans?

2. You have 10 beans but you want 12 beans. How many more must you get?

3. 10 beans plus 5 beans equals how many beans?

4. You have 10 beans but want 19 beans. How many more must you get?

5. 10 beans added to 1 bean make how many beans?

6. 14 beans minus 10 beans equals how many beans?

7. 10 beans added to 7 beans make how many beans?

8. You have 10 beans and want 20 beans. How many more must you get?

9. 10 beans added to 4 beans make how many beans?

10. You have 10 beans but want 16 beans. How many more beans do you need?

11. 10 beans added to 9 beans make how many beans?

12. 11 beans take away 10 beans leaves how many beans?

13. 10 beans plus 10 beans equals how many beans?

14. You have 17 beans and take 10 away. How many beans do you have?

15. 10 beans added to 8 beans make how many beans?

16. 18 beans minus 10 beans equals how many beans?

17. 10 beans added to 6 beans make how many beans?

18. 18 beans take away 10 beans leaves how many beans?

19. 10 beans and 2 beans together make how many beans?

20. 13 beans take away 10 beans leaves you with how many beans?

Answers

1. 13 beans

Tip: Be sure to require fully worded answers.

2. 2 beans
3. 15 beans
4. 9 beans
5. 11 beans
6. 4 beans
7. 17 beans
8. 10 beans
9. 14 beans
10. 6 beans
11. 19 beans
12. 1 bean
13. 20 beans
14. 7 beans
15. 18 beans
16. 8 beans
17. 16 beans
18. 8 beans
19. 12 beans
20. 3 beans

Step 6: Give mixed sums with imaginary objects.

Tip: On days when the work is going exceedingly well, you may have your child write on his slate or in his math notebook an addition or subtraction equation he has solved, using the proper signs.

1. 10 flowers and 10 flowers?

2. 10 cookies from 11 cookies?

3. 10 pictures from 14 pictures?

4. 10 teeth and 4 teeth?

5. 10 pens and 2 pens?

6. 10 magnets and 1 magnet?

7. 10 paper clips and 8 paper clips?

8. 10 washcloths from 16 washcloths?

9. 10 books from 17 books?

10. 10 hats and 3 hats?

Answers

1. 20 flowers

Tip: Be sure to require fully worded answers.

2. 1 cookie
3. 4 pictures
4. 14 teeth
5. 12 pens
6. 11 magnets
7. 18 paper clips
8. 6 washcloths
9. 7 books
10. 13 hats

Step 7: Present mixed sums using pure number.

Tip: Be sure to put away the manipulatives when they are no longer needed for this table.

How much is left if you take

1. 10 from 12?

How much have you if you add

6. 10 and 4?

2. 10 from 18?

3. 10 from 13?

4. 10 from 10?

5. 10 from 14?

7. 10 and 5?

8. 10 and 7?

9. 10 and 9?

10. 10 and 2?

Answers

Tip: Be sure to require fully worded answers.

1. 2
2. 8
3. 3
4. 0
5. 4
6. 14
7. 15
8. 17
9. 19
10. 12

Review 10 and Previous Tables: Addition

1. 4 fire trucks pass by with 2 ambulances. How many vehicles passed by in all?

2. Christopher finds 6 snakes and the next day finds 1 more. How many snakes did he find all together?

3. There are 10 paving stones along one side of the driveway and 10 more along the other side. How many stones line the driveway?

4. 5 sailboats and 5 motorboats are on the lake. How many boats in all?

5. Ruth swam 6 laps at the pool yesterday and 4 laps today. How many laps did she swim in all?

6. 4¢ + 5¢ =

7. 6¢ + 9¢ =

8. 10¢ + 2¢ =

9. 5¢ + 8¢ =

10. 7¢ + 2¢ =

11.

6 bunnies	6 sparrows	5 feathers	8 pillows	9 acorns
+ 5 bunnies	+ 8 sparrows	+ 9 feathers	+ 4 pillows	+ 10 acorns

12. 10 + 4

13. 2 + 8

14. 3 + 7

15. 6 + 3

16. 8 + 8

Answers

1. 6 vehicles

Tip: Be sure to require fully worded answers.

2. 7 snakes
3. 20 stones
4. 10 boats
5. 10 laps
6. 9¢
7. 15¢
8. 12¢
9. 13¢
10. 9¢
11.

6 bunnies	6 sparrows	5 feathers	8 pillows	9 acorns
+ 5 bunnies	+ 8 sparrows	+ 9 feathers	+ 4 pillows	+ 10 acorns
11 bunnies	14 sparrows	14 feathers	12 pillows	19 acorns

12. 14

13. 10
14. 10
15. 9
16. 16

Review 10 and Previous Tables: Subtraction

1. Jeremiah had $20. He bought a watch for $10. How much does he have now?

2. Ren had 19 marbles and lost 9. How many marbles remain?

3. Shepherd was given 17¢ and spent 9¢ at the fair. How much does he have left?

4. Ryan made 14 postcards. If he sends 4 of them in the mail, how many postcards does he have?

5. 9 pupils are absent from a class of 16. How many pupils are in class?

6. 19¢ – 10¢ =

7. 13¢ – 4¢ =

8. 13¢ – 9¢ =

9. 18¢ – 9¢ =

10. 14¢ – 6¢ =

11.

16 bears	11 lizards	17 pinecones	15 forks	13 flowers
– 9 bears	– 4 lizards	– 10 pinecones	– 9 forks	– 10 flowers

12. 9 – 7

13. 18 – 10

14. 12 – 2

15. 12 – 8

16. 14 – 10

Answers

1. $10

Tip: Be sure to require fully worded answers.

2. 10 marbles
3. 8¢
4. 10 postcards
5. 7 pupils
6. 9¢
7. 9¢
8. 4¢
9. 9¢
10. 8¢
11.

16 bears	11 lizards	17 pinecones	15 forks	13 flowers
– 9 bears	– 4 lizards	– 10 pinecones	– 9 forks	– 10 flowers
7 bears	7 lizards	7 pinecones	6 forks	3 flowers

12. 2
13. 8
14. 10
15. 4
16. 4

If your child needs more practice to be comfortable with this table, you can find additional review sums for 10 and previous tables on page 283. Also look at previous lessons for many more sums that you can use as is or by changing names, numbers, or objects.

Addition with Larger Numbers

With the knowledge of the addition and subtraction tables, we are ready to help further our student's understanding of addition. We will begin again in the concrete, using money. Although this will initially be a review from Book 1, be sure to allow the student the time needed to work the problems and continue to encourage fully worded answers.

For any answers that total 10 or above, be sure to have your child exchange the pennies for dimes and pennies as an aid in understanding place value. Changing the pennies to dimes also forms a natural bridge when sums involving "carrying" are introduced.

As the student progresses with the concepts, he will transition naturally to larger numbers using pennies, dimes, and dollars. Students will be working with numbers through 1,000.

Keep lessons to 15 minutes of work on the newest concept and 5 minutes of review at the end, making each day's math time 20 minutes total. Simply set the timer for 15 minutes and work on the lessons as given. When the 15 minutes are up, spend 5 minutes reviewing past math concepts. Review sections are included in the lessons. You will find more review sums (for both addition and subtraction) in the back of this book, beginning on page 285, or feel free to reuse sums from previous lessons, changing names, numbers, and objects as desired.

Before beginning these lessons, you may wish to view the demonstrations in the *Charlotte Mason's Living Math: A Guided Journey* session on "Addition with Larger Numbers" to familiarize yourself further with the methods.

Review Addition with Smaller Numbers

Tip: As a help, pull a Number Sentence Card out before the lesson begins so your student's work can continue uninterrupted if you get called away for a moment.

Small Sums with Some Exchanging

Objects used: coins

1. 4 pennies + 3 pennies =

2. 7 pennies + 2 pennies =

3. 8 pennies + 3 pennies =

Tip: For any answers that total 10 or more, be sure to have your child exchange the pennies for dimes and pennies as an aid in understanding place value; i.e., 8 + 3 = 11 pennies, exchange for 1 dime and 1 penny.

4. 5 pennies + 6 pennies =

5. 9 pennies + 4 pennies =

6. 1 dime + 7 pennies =

7. 4 pennies + 8 pennies =

8. 3 pennies + 10 pennies =

9. 7 pennies + 8 pennies =

10. 5 pennies + 5 pennies = *To be written after solving.*

11. 10 pennies + 5 pennies =

12. 8 pennies + 9 pennies =

13. 1 penny + 4 pennies =

14. 10 pennies + 4 pennies =

15. 1 penny + 1 penny =

16. 3 pennies + 5 pennies =

17. 7 pennies + 3 pennies =

18. 6 pennies + 1 penny =

19. 5 pennies + 2 pennies =

20. 4 pennies + 4 pennies = *To be written after solving.*

Answers

1. 7 pennies

Tip: Be sure to require fully worded answers; i.e., 4 + 3 = 7.

2. 9 pennies
3. 11 pennies, exchange for 1 dime and 1 penny
4. 11 pennies, exchange for 1 dime and 1 penny
5. 13 pennies, exchange for 1 dime and 3 pennies
6. 8 pennies
7. 12 pennies, exchange for 1 dime and 2 pennies
8. 13 pennies, exchange for 1 dime and 3 pennies
9. 15 pennies, exchange for 1 dime and 5 pennies
10. 10 pennies, exchange for 1 dime.
11. 15 pennies, exchange for 1 dime and 5 pennies
12. 17 pennies, exchange for 1 dime and 7 pennies
13. 5 pennies
14. 14 pennies, exchange for 1 dime and 4 pennies
15. 2 pennies

16. 8 pennies
17. 10 pennies, exchange for 1 dime
18. 7 pennies
19. 7 pennies
20. 8 pennies

Longer Sums with and without Exchanging

Objects used: coins

Tip: Be sure to always have your child exchange the pennies into a dime or dime and pennies when applicable. Allow plenty of time for him to accomplish the work.

1. 1 penny + 2 pennies + 6 pennies =

2. 4 pennies + 5 pennies + 2 pennies =

3. 5 pennies + 5 pennies + 9 pennies =

4. 3 pennies + 1 penny + 4 pennies =

5. 6 pennies + 2 pennies + 1 penny =

6. 1 penny + 2 pennies + 4 pennies =

7. 9 pennies + 1 penny + 9 pennies =

8. 5 pennies + 6 pennies + 2 pennies =

9. 8 pennies + 3 pennies + 4 pennies =

10. 9 pennies + 4 pennies + 2 pennies =

11. 2 pennies + 7 pennies + 3 pennies =

12. 6 pennies + 5 pennies + 5 pennies =

13. 8 pennies + 4 pennies + 1 penny =

14. 4 pennies + 5 pennies + 4 pennies =

15. 2 pennies + 4 pennies + 3 pennies =

16. 9 pennies + 2 pennies + 5 pennies =

17. 5 pennies + 1 penny + 2 pennies =

18. 8 pennies + 3 pennies + 2 pennies =

19. 4 pennies + 2 pennies + 4 pennies =

20. 9 pennies + 4 pennies + 6 pennies = *To be written after solving.*

Answers

1. 9 pennies

Tip: Be sure to require fully worded answers

2. 11 pennies, exchange for 1 dime and 1 penny
3. 19 pennies, exchange for 1 dime and 9 pennies
4. 8 pennies
5. 9 pennies
6. 7 pennies
7. 19 pennies, exchange for 1 dime and 9 pennies
8. 13 pennies, exchange for 1 dime and 3 pennies
9. 15 pennies, exchange for 1 dime and 5 pennies
10. 15 pennies, exchange for 1 dime and 5 pennies
11. 12 pennies, exchange for 1 dime and 2 pennies
12. 16 pennies, exchange for 1 dime and 6 pennies
13. 13 pennies, exchange for 1 dime and 3 pennies
14. 13 pennies, exchange for 1 dime and 3 pennies
15. 9 pennies
16. 16 pennies, exchange for 1 dime and 6 pennies
17. 8 pennies

18. 13 pennies, exchange for 1 dime and 3 pennies
19. 10 pennies, exchange for 1 dime
20. 19 pennies, exchange for 1 dime and 9 pennies

Review

1. 8 birds at the bird feeder when 4 flew off. How many birds remained at the feeder?

2. Luke had 16¢ then gave his sister 10¢. How many cents did Luke still have?

3. Aviyah made 12 cookies and 5 were eaten. How many cookies remained?

4. Peter gathered 18 blackberries. If he gave 9 to his sister, how many had he left?

5. Corin wants to read 15 pages in her book before bed. She has read 4. How many has she left?

6. 11¢ – 6¢ =

7. 14¢ – 3¢ =

8. 20¢ – 10¢ =

9. 7¢ – 3¢ =

10. 17¢ – 6¢ =

11.

12 cardinals	5 sparrows	8 juncos	9 wrens	6 bluejays
– 2 cardinals	– 2 sparrows	– 6 juncos	– 4 wrens	– 3 bluejays

12. 13 – 3

13. 9 – 2

14. 10 – 4

15. 14 – 10

16. 11 – 2

Answers

1. 4 birds

Tip: Be sure to require fully worded answers

2. 6¢
3. 7 cookies
4. 9 blackberries
5. 11 pages
6. 5¢
7. 11¢
8. 10¢
9. 4¢
10. 11¢
11.

12 cardinals	5 sparrows	8 juncos	9 wrens	6 bluejays
– 2 cardinals	– 2 sparrows	– 6 juncos	– 4 wrens	– 3 bluejays
10 cardinals	3 sparrows	2 juncos	5 wrens	3 bluejays

12. 10
13. 7
14. 6
15. 4
16. 9

Review Numeration and Notation to 100

Pennies and Dimes and Counting by Tens to 100

(instructions to the parent)

Objects used: coins

1. Ask your child how many pennies make a dime. (10)

2. Have your child take out 10 dimes and see if he can count by 10s to 100. If he is able to do so, skip to the Notation section of this lesson to review writing by 10s to 100. If he is not able to count by 10s to 100, continue in this section for a step-by-step review.

3. Ask your child: How many pennies make 2 dimes? This is a review from Book 1, but if this process is unclear you may say something such as: You know there are 10 pennies in a dime. The word *ten* comes from the Latin for *ten* which is *tig*. Just as our next number after *9* is *10*, our next number after *19* is *20*. The word *twenty* comes from Latin as well, with *twen* meaning *two* and *tig* again, meaning *ten*; so *twenty* means "two tens." When we have 2 dimes we have two tens.

4. Now ask: If twenty pennies make 2 dimes, how many pennies are contained in 3 dimes? (30) 4 dimes? (40)

5. Continue asking how many pennies are contained in 5 dimes, etc., up to how many pennies make 9 dimes. (50, 60, 70, 80, 90)

6. Ask: How many pennies would 10 dimes make? (100)

7. Now you may say something such as: Just as when you had 10 pennies and exchanged them for a dime, what do you suppose we could do with 10 dimes? As this is a review, your child should be able to answer something to the effect of "Exchange it for a dollar." If not,

introduce the dollar, saying something like: We have a piece of paper money called one dollar and it is worth 100 pennies or ten dimes.

Notation by Tens to 100 *(instructions to the parent)*

Objects used: dimes

1. Have your child get out 1 dime.

2. Draw *10* on the gridded slate or graph paper in order to show the written figure for *ten*, with the *0* standing for no units and the *1* in the tens place to show one ten. Write one numeral per square. Be sure to leave space for the hundreds place on the left.

3. Point to the *0* and ask the child how many units, and then point to the *1* and ask how many tens.

4. Now it is the child's turn to write the number *10* on the slate.

5. Finally, have the child write the number *10* in his math notebook.

6. Continue steps 1–5 until the child has written *10, 20, 30,* etc. through *100* in a column, first on the slate and then in his math notebook, being sure to keep proper place value.

Review

1. Gabe had 8¢ and put 5¢ in his piggy bank. How much had he left out?

2. Hudson had 17¢ and gave 7¢ to his sister. How much had he left?

Tip: If any coins are gotten out during this review, be sure to use a dime and pennies for anything over 10.

3. Annalise had 12¢. After spending 5¢ on candy, how much had she?

4. Lyla had 10¢ and spent 5¢. How much had she left?

5. Wyatt had 10¢ and spent 3¢. How much had he left?

6. Simon had 11¢ and spent 3¢. How much had he left?

7. Abby had 15¢ and put 10¢ in her piggy bank. How much had she left out?

8. If you have 16¢ and spend 8¢, how much have you remaining?

9. If you have 8¢ and spend 3¢, how much remains?

10. If you have 9¢ and spend 8¢, how much have you?

11. If you have 14¢ and spend 5¢, how much remains?

12. If you have 15¢ and spend 7¢, how much have you left?

13. If you have 6¢ and spend 6¢, how much remains?

14. If you have 20¢ and spend 20¢, how much have you left?

15. If you have 7¢ and spend 5¢, how much have you?

16. If you have 13¢ and spend 6¢, how much have you remaining?

17. If you have 4¢ and spend half, how much have you?

Tip: To be taken experimentally.

18. If you have 18¢ and spend half, how much have you?

19. If you have 11¢ and spend 4¢, how much have you remaining?

20. If you have 5¢ and spend 1¢, how much remains?

Answers

1. 3¢

Tip: Be sure to require fully worded answers

2. 10¢
3. 7¢
4. 5¢
5. 7¢
6. 8¢
7. 5¢
8. 8¢
9. 5¢
10. 1¢
11. 9¢
12. 8¢
13. 0¢
14. 0¢
15. 2¢
16. 7¢
17. 2¢
18. 9¢
19. 7¢
20. 4¢

Addition of Double Digits without Carrying

Tip: Keep lessons to 15 minutes of work on the newest concept and 5 minutes of review at the end, making each day's math time 20 minutes total.

Now we will progress to using larger denominations with the addition of dimes. Be sure your child always keeps dimes on his left and pennies on his right. Exchanging the pennies for dimes forms a natural bridge when sums involving "carrying" are introduced. The child will be working a number of sums here —not so many that he becomes bored but enough to ensure and cement his understanding.

Have him speak aloud as he solves the problem. When the problems are worked with ease, the student may write a sum or two in his math notebook before moving on. This may be the first time he is writing problems in a column, rather than a sentence format, so a detailed sample is provided.

Sample *(instructions to the parent)*

Objects used: coins

Read aloud the problem below. (Do not read the answer.) Have your student speak the problem aloud as he solves it with coins.

  3 dimes and 2 pennies
+ 4 dimes and 5 pennies
  7 dimes and 7 pennies, or 77¢

Ask if the student knows how to write the equation with the proper signs in a column. If not, show him how to write the problem (see below), using a sheet of gridded paper or a slate. Place each numeral and symbol in its own square of the grid. Explain that the line under the *45* means *equals* or *is equal to*.

$$\begin{array}{r} 32 \\ +45 \\ \hline 77 \end{array}$$

Sums

Now give your child some or all of the equations below orally. The student should speak through the process and solve them with coins. Have him write a few of the problems either while solving or after solving, as demonstrated above.

1. 3 dimes and 7 pennies
   + 4 dimes and 2 pennies

2. 7 dimes and 5 pennies
   + 2 dimes and 4 pennies

Tip: When working with columns, encourage the student to get in the habit of beginning to solve the problem with the pennies, or units, first.

3. 1 dime and 8 pennies
+ 5 dimes and 1 penny

4. 1 dime and 3 pennies
+ 2 dimes and 5 pennies

5. 7 dimes and 2 pennies
+ 1 dime and 6 pennies

6. 6 dimes and 4 pennies
+ 1 dime and 4 pennies

7 3 dimes and 2 pennies
+ 1 dime and 4 pennies

8. 8 dimes and 5 pennies
+ 1 dime and 4 pennies

9. 2 dimes and 4 pennies
+ 4 dimes and 2 pennies

10. 5 dimes and 4 pennies
+ 3 dimes and 2 pennies

Answers

1. 7 dimes and 9 pennies, or 79¢

Tip: Be sure to require the student to speak aloud as he solves the equations and to give both parts of the answer: the number of dimes and pennies and the total amount of money.

2. 9 dimes and 9 pennies, or 99¢
3. 6 dimes and 9 pennies, or 69¢
4. 3 dimes and 8 pennies, or 38¢
5. 8 dimes and 8 pennies, or 88¢
6. 7 dimes and 8 pennies, or 78¢
7. 4 dimes and 6 pennies, or 46¢
8. 9 dimes and 9 pennies, or 99¢
9. 6 dimes and 6 pennies, or 66¢
10. 8 dimes and 6 pennies, or 86¢

Addition of Double Digits with Carrying

Sample *(instructions to the parent)*
Objects used: coins

Read aloud the equation below.

	4 dimes	and	3 pennies
+	2 dimes	and	8 pennies

Write the equation on the slate or grid paper.

$$\begin{array}{r} 43 \\ +28 \\ \hline \end{array}$$

Have your student work the equation with coins, using dimes and pennies. It will be helpful to have him speak the process aloud as he works and exchanges pennies for dimes. (For example, 3 pennies plus 8 pennies are 11 pennies, which make 1 dime and 1 penny. 1 dime plus 4 dimes plus 2 dimes make 7 dimes.)

	1 dime		
	4 dimes	and	3 pennies
+	2 dimes	and	8 pennies
	7 dimes		1 penny

Tip: Encourage the student to get in the habit of starting with the pennies, or units.

Now show the student how to notate that process in writing. If desired, see if your child can tell you how it is done. If not, simply show and tell him, and he can talk through the next problem. (For example, 3 pennies plus 8 pennies are 11 pennies, which make 1 dime and 1 penny. The 1 penny is written in the units place and the 1 dime is written in the tens place. 1 dime plus 4 dimes plus 2 dimes are 7 dimes. The 7 is written in the tens place.)

$$\begin{array}{r} {}^{1} \\ 43 \\ +28 \\ \hline 71 \end{array}$$

Sums

1. 5 dimes and 4 pennies
 \+ 2 dimes and 6 pennies

2. 1 dime and 8 pennies
 \+ 3 dimes and 2 pennies

3. 4 dimes and 5 pennies
 \+ 3 dimes and 7 pennies

4. 6 dimes and 4 pennies
 \+ 2 dimes and 9 pennies

5. 7 dimes and 9 pennies
 \+ 2 pennies

6. 2 dimes and 8 pennies
 \+ 1 dime and 3 pennies

7. 3 dimes and 6 pennies
 \+ 3 dimes and 7 pennies

8. 2 dimes and 8 pennies
 \+ 6 dimes and 4 pennies

Tip: Have your student write out one or two of the following equations in his math notebook, either as it is solved or after it is solved.

9. 2 dimes and 5 pennies
 \+ 3 dimes and 6 pennies

10. 2 dimes and 7 pennies
 \+ 5 dimes and 4 pennies

9. Written:

$$\begin{array}{r} {}^{1} \\ 25 \\ +\ 36 \\ \hline 61 \end{array}$$

10. Written:

$$\begin{array}{r} {}^{1} \\ 27 \\ +\ 54 \\ \hline 81 \end{array}$$

Answers

1. 8 dimes and 0 pennies, or 80¢

Tip: Be sure to require the student to speak aloud as he solves the equations.

2. 5 dimes and 0 pennies, or 50¢
3. 8 dimes and 2 pennies, or 82¢
4. 9 dimes and 3 pennies, or 93¢
5. 8 dimes and 1 penny, or 81¢
6. 4 dimes and 1 penny, or 41¢
7. 7 dimes and 3 pennies, or 73¢
8. 9 dimes and 2 pennies, or 92¢
9. 6 dimes and 1 penny, or 61¢
10. 8 dimes and 1 penny, or 81¢

Mixed Practice with and without Carrying

Tip: Keep lessons to 15 minutes of work on the newest concept and 5 minutes of review at the end, making each day's math time 20 minutes total.

Sums

Objects used: coins

Tip: Use as many of these problems as needed to solidify understanding.

1. 1 dime and 5 pennies + 4 dimes and 3 pennies	2. 3 dimes and 2 pennies + 4 dimes and 7 pennies

Tip: Help your student get in the habit of starting with the pennies, or units.

3. 4 dimes and 6 pennies + 3 dimes and 6 pennies	4. 2 dimes and 8 pennies + 2 dimes and 2 pennies
5. 3 dimes and 7 pennies + 2 dimes and 1 penny	6. 1 dime and 5 pennies + 3 dimes and 5 pennies
7. 4 dimes and 6 pennies + 1 dime and 3 pennies	8. 5 dimes and 9 pennies + 2 pennies

Tip: You may have your student write one or two of the following sums in his math notebook if work is going exceptionally well.

9. 2 dimes and 8 pennies
 + 1 dime and 4 pennies

10. 5 dimes and 7 pennies
 + 2 dimes and 1 penny

9. Written:

$$\begin{array}{r} {}^{1} \\ 28 \\ +\ 14 \\ \hline 42 \end{array}$$

10. Written:

$$\begin{array}{r} 57 \\ +\ 21 \\ \hline 78 \end{array}$$

Answers

1. 5 dimes and 8 pennies, or 58¢

Tip: Encourage your child to speak the process aloud as he solves the equations.

2. 7 dimes and 9 pennies, or 79¢
3. 8 dimes and 2 pennies, or 82¢
4. 5 dimes and 0 pennies, or 50¢
5. 5 dimes and 8 pennies, or 58¢
6. 5 dimes and 0 pennies, or 50¢
7. 5 dimes and 9 pennies, or 59¢
8. 4 dimes and 1 penny, or 41¢
9. 4 dimes and 2 pennies, or 42¢
10. 7 dimes and 8 pennies, or 78¢

Review

1. Dakota is 11 and her brother is 6. How many years older is Dakota than her brother?

2. Levi must pick 12 lemons. He has picked 8. How many more must he pick?

3. There are 5 eggs in a nesting box but 2 broke. How many eggs remain whole?

4. Kaden has 7¢ and Sunny has 7¢. How much have they together?

5. Macy had 2 books about horses and received 4 more on her birthday. How many books has she about horses now?

6. What is 9 plus 3?

7. What is 2 plus 5?

8. What is 4 taken from 7?

9. What is 3 added to 7?

10. What is 9 minus 4?

Answers

1. 5 years

Tip: Always require fully worded answers.

2. 4 lemons
3. 3 eggs
4. 14¢
5. 6 books
6. 12
7. 7
8. 3
9. 10
10. 5

Hundreds through One Thousand

(instructions to the parent)

1. As a small form of review, which should lead the child to think through the task at hand, ask: How many pennies are in a dime? (10)

2. Now ask: How many pennies are in 2 dimes? (20)

3. Next, see if he remembers how many pennies are in a dollar. (100)

4. Write *100* on a slate or gridded paper, leaving room below for more numbers as well as space to the left, which will later hold the thousands' place.

5. Ask: If 100 pennies are in a dollar, how many pennies are in 2 dollars? (200)

6. Write *200* below the *100.*

7. Point to the first zero on the right and ask: How many units? (0) Point to the zero in the middle and ask: How many tens? (0) Point to the *2* and ask: How many hundreds? (2)

8. Say: Just as we would count "97, 98, 99, 100," we also count to 200 in the same way, saying, "197, 198, 199, 200." If your student knows, allow him to finish the counting.

Tip: Be sure to say the three-digit numbers as "one hundred ninety-seven, one hundred ninety-eight, etc." rather than "one ninety-seven."

9. Allowing time for your child to think, ask: If 1 dollar contains 100 pennies and 2 dollars contain 200 pennies, how many pennies are contained in 3 dollars? (300)

10. Once answered, you will write *300* below the *200.*

11. Continue steps 9 and 10 up through *900.* Every so often you may point to a number and ask how many hundreds, units, or tens are contained in it, in any order.

12. Now, pointing to the numbers in order, have your child read through: *one hundred, two hundred, three hundred, four hundred, five hundred, six hundred, seven hundred, eight hundred, nine hundred.*

13. Ask if your child knows what would come after *900* if counting by hundreds. He may say *ten hundred;* but if not, tell him it is *ten hundred,* which we also call *one thousand.*

14. Show him how to write *1000* by writing it under *900*, being sure to keep proper place value.

15. Now, have your child copy the numbers neatly into his math notebook.

Exchanging Dimes and Dollars

Sums

Objects used: dollars, dimes, pennies

1. If you have 10 dimes but don't want to carry so much change, what could you exchange it for?

Tip: You may use play dollars, dollars from a board game, or real dollars.

2. If you have 12 dimes but don't want to carry so much change, what could you exchange it for?

3. Show me $2.30 in dollars and dimes.

4. If you saved 500 pennies in a jar, how many dollars could you exchange them for?

5. If you have 16 dimes, how many dollars could you exchange them for and how many dimes remain?

6. Show me $3.41 in dollars, dimes, and pennies.

7. If you had $3.41 all in pennies, how many pennies would you have? *(To be answered orally without manipulatives.)*

8. If you have 162 pennies, how many dollars and dimes could you exchange them for?

9. Show me $5.20 in dollars and dimes.

10. Exchange 11 dimes for dollars and dimes.

Answers

1. One dollar

Tip: Be sure to require fully worded answers; i.e., ten dimes equal one dollar.

2. One dollar and two dimes
3. 2 dollars, 3 dimes, zero pennies
4. 5 dollars
5. 1 dollar and 6 dimes
6. 3 dollars, 4 dimes, 1 penny
7. 341 pennies
8. 1 dollar, 6 dimes, 2 pennies
9. 5 dollars, 2 dimes, no pennies
10. 1 dollar and 1 dime

Review

1. Two dogs and 1 puppy are in the yard. How many tails in all?

2. Two girls and a dog are playing with a ball. How many legs in all?

3. Marianne ate 9 cherries and Elinor ate 6. How many more cherries did Marianne eat than Elinor?

4. Jack checked out 3 books and Peter checked out 5. How many books had they checked out together?

5. Sunny checked out 8 books from the library and her brother checked out 4. How many more books had Sunny checked out than her brother?

6. What is 7 added to 3?

7. What is 5 from 9?

8. What is 6 plus 8?

9. What is 4 added to 2?

10. What is 10 minus 8?

Answers

1. 3 tails
2. 8 legs
3. 3 cherries
4. 8 books
5. 4 books
6. 10
7. 4
8. 14
9. 6
10. 2

Addition Using Dollars without Carrying

Tip: As a help, pull a Number Sentence Card out before the lesson begins so your student's work can continue uninterrupted if you get called away for a moment.

Sample *(instructions to the parent)*

Objects used: dollars, dimes, pennies

Read aloud the equation below and have your student speak through the process as he solves it with dollars, dimes, and pennies. Help your student get in the habit of starting with the pennies, or units (for example, 4 pennies plus 5 pennies are 9 pennies, 3 dimes plus 2 dimes are 5 dimes, 2 dollars plus 2 dollars are 4 dollars).

```
 $2.34
+2.25
```

Tip: Be sure dollars are kept on the student's left, followed by dimes and pennies, all in their proper places.

Now show your student how the equation is written (see below). At this point you need only say that $ stands for dollars and the dot "." is placed between the dollars and cents. Decimals will be formally introduced and more fully explored in subsequent books.

```
 $2.34
+2.25
 $4.59
```

Sums

Tip: You may be surprised by how many numbers a child using Charlotte Mason's approach to arithmetic is able to keep in his head, but if need be, feel free to write the numbers on a slate or graph paper.

```
1.  $6.22    2.  $4.64    3.  $3.81    4.  $2.43    5.  $1.35
    +3.41        +1.23        +  .11       +5.04        +2.32

6.  $3.43    7.  $2.21    8.  $4.13    9.  $7.33    10. $5.61
    +1.42        +3.54        +  .76       +1.25        +4.34
```

Answers

```
1.  $6.22    2.  $4.64    3.  $3.81    4.  $2.43    5.  $1.35
    +3.41        +1.23        +  .11       +5.04        +2.32
    $9.63        $5.87        $3.92        $7.47        $3.67
```

Tip: The student should speak aloud as he solves the equations. At the end of his explanation, he may say only the answer without restating the entire problem.

6. \$3.43	7. \$2.21	8. \$ .76	9. \$7.33	10. \$5.61
+1.42	+3.54	+4.13	+1.25	+4.34
\$4.85	\$5.75	\$4.89	\$8.58	\$9.95

Addition Using Dollars with Carrying

Tip: Keep lessons to 15 minutes of work on the newest concept and 5 minutes of review at the end, making each day's math time 20 minutes total.

Samples *(instructions to the parent)*

Objects used: dollars, dimes, pennies

Two sample problems will be written out by the teacher to show how problems of this type are written, while the student solves the problems by using dollars, dimes, and pennies—exchanging pennies for dimes and dimes for dollars as necessary.

Tip: Be sure dollars are kept on the student's left, followed by dimes and pennies to his right.

1. Write this equation on the slate or grid paper.

   \$5.67
   +3.84

   The child places 5 dollars, 6 dimes, and 7 pennies on the table then places 3 dollars, 8 dimes, and 4 pennies below it. He should start solving by adding the two units together, speaking through the process (i.e., 7 pennies plus 4 pennies make 11 pennies. That means

1 dime, and 1 penny left over). Show how to notate that step on the equation:

$$\begin{array}{r} 1 \\ \$5.67 \\ \underline{+3.84} \\ 1 \end{array}$$

Tip: Help your student get in the habit of starting with the pennies, or units.

Now have the student talk through the process of adding the dimes (i.e., 1 dime and 6 dimes and 8 dimes = 15 dimes. That is 1 dollar and leaves 5 dimes over). Show him how to notate that step.

$$\begin{array}{r} 1\ \ 1 \\ \$5.67 \\ \underline{+3.84} \\ 51 \end{array}$$

Last, have the student add the dollars, talking through that step and giving the final answer (i.e., 1 dollar and 5 dollars and 3 dollars make 9 dollars. The answer is \$9.51). Show how to notate that step.

$$\begin{array}{r} 1\ \ 1 \\ \$5.67 \\ \underline{+3.84} \\ \$9.51 \end{array}$$

2. Do another sample equation, allowing the child to notate as much as he is able. Be careful, though, that the writing does not overshadow his thinking process.

   Write the equation. (The child gets out 4 dollars, 1 dime, and 8 pennies then places below it 2 dollars, 9 dimes, and 6 pennies.)

$$\begin{array}{r} \$4.18 \\ \underline{+2.96} \end{array}$$

   Have the student speak and show how to add the units, or pennies, first. (8 pennies plus 6 pennies make 14 pennies. That is 1 dime and 4 pennies.)

$$\begin{array}{r} {\scriptstyle 1} \\ \$4.18 \\ \underline{+2.96} \\ 4 \end{array}$$

Next do the tens column, or dimes. (1 dime plus 1 dime plus 9 dimes make 11 dimes. That is 1 dollar and 1 dime left over.)

$$\begin{array}{r} {\scriptstyle 1\ 1} \\ \$4.18 \\ \underline{+2.96} \\ 14 \end{array}$$

Last, do the hundreds column, or dollars. (1 dollar plus 4 dollars plus 2 dollars equals 7 dollars.)

$$\begin{array}{r} {\scriptstyle 1\ 1} \\ \$4.18 \\ \underline{+2.96} \\ \$7.14 \end{array}$$

Sums

Tip: The student should speak aloud as he solves the equations.

1. $\begin{array}{r} \$2.63 \\ \underline{+4.85} \end{array}$	2. $\begin{array}{r} \$5.63 \\ \underline{+1.67} \end{array}$	3. $\begin{array}{r} \$4.05 \\ \underline{+2.69} \end{array}$	4. $\begin{array}{r} \$5.90 \\ \underline{+3.25} \end{array}$	5. $\begin{array}{r} \$6.86 \\ \underline{+1.54} \end{array}$
6. $\begin{array}{r} \$1.42 \\ \underline{+2.63} \end{array}$	7. $\begin{array}{r} \$4.55 \\ \underline{+\ \ .64} \end{array}$	8. $\begin{array}{r} \$2.10 \\ \underline{+\ \ .97} \end{array}$	9. $\begin{array}{r} \$3.65 \\ \underline{+1.78} \end{array}$	10. $\begin{array}{r} \$4.58 \\ \underline{+5.13} \end{array}$

Tip: Your student may write the final equation on each line in his math notebook after solving (i.e., #5 and #10).

Answers

1	1 1	1	1	1 1
1. $2.63	2. $5.63	3. $4.05	4. $5.90	5. $6.86
+4.85	+1.67	+2.69	+3.25	+1.54
$7.48	$7.30	$6.74	$9.15	$8.40

1	1	1	1 1	1
6. $1.42	7. $4.55	8. $2.10	9. $3.65	10. $4.58
+2.63	+ .64	+ .97	+1.78	+5.13
$4.05	$5.19	$3.07	$5.43	$9.71

Review

1. Silas and Jonah each picked 8 apples. How many have they picked in all?

2. There are 9 red balloons and 9 white balloons. How many balloons in all?

3. There were 12 balloons but 6 flew away. How many balloons remained?

4. There were 10 sheep and 7 goats in the farmyard. How many more sheep than goats?

5. There are 2 sheep and a lamb in a meadow. How many legs in all?

6. What is 3 taken from 9?

7. What is 10 minus 2?

8. What is 5 added to 6?

9. What is 8 minus 3?

10. What is 7 minus 3?

Answers

1. 16 apples
2. 18 balloons
3. 6 balloons
4. 3 sheep
5. 12 legs
6. 6
7. 8
8. 11
9. 5
10. 4

Numeration and Notation to 1000

(instructions to the parent)

1. Write *6* on a slate, being sure to leave two places to the left, and ask your child to read: 6.

2. Place a *5* in front of it to make *56,* and have your child read: 56.

3. Now place a *3* in front to make *356*, and ask your child to read: 356.

Tip: Make sure he says three hundred fifty-six, *rather than just* three fifty-six.

4. If he is able to read that number easily, skip to step 5. If he is unable to read the number, explain as follows: When three numbers are written side by side, such as *356*, the first on the right represents units, the next to the left of it shows tens, and the third represents hundreds. Together it is read *three hundred fifty-six.*

5. Write on the slate and have your child read: *528, 217, 221, 768, 444, 111, 959, 693, 737, 777, 810, 545, 143, 999.*

6. Write *768* and ask: How many units? tens? and hundreds?

7. Next, have your child write on a slate or gridded paper *two hundred fifty-four* in numerals, that is *254*. If he is unsure, you may direct him by saying something like this: First, write the number that represents the hundreds. Then which number will represent the tens? We write *5* to show the 5 tens. Which number to represent the units? We write *4* to represent the 4 units. If further instruction is still necessary, you may say: 254. First, write the *2*, representing the hundreds, then the *54*.

8. Now, going slowly enough and encouraging neatness, have your child write the following numbers in figures: *eight hundred twenty-three; three hundred seventeen; seven hundred sixty-one; four hundred thirty-nine; one hundred fifty-four.*

Tip: Be careful not to overtax your child's attention with reading and writing in this lesson. If he is beginning to weary, even if the 15 minutes are not up, move on to doing some review mental arithmetic from previous lessons or the extra sums in the back of this book. You can finish this lesson next time if needed. Better to secure full attention and a cheerful spirit in his math lesson than to check off a page as done.

9. Write on a slate and have your child read: *400, 800, 500, 700, 200, 1000.* This should be clear but, if not, explain: There are no units or tens in *400,* so we read simply *four hundred,* etc.

10. Ask: How many units in 400? How many hundreds? How many tens?

11. Ask: How many units make 1 ten? How many tens make 1 hundred? How many hundreds make 1 thousand?

12. Ask your child to write on his slate in numerals: *five hundred; two hundred; nine hundred; three hundred; one thousand.* If he needs direction you may say: To show hundreds, we must always have three places. We write a *5* to show the five hundreds, and then a *0* in the tens place since there are no tens, and a *0* in the units place to show there are no units. If necessary: To show thousands then, we must

always have four places. We write a *1* to show the thousand, and then a *0* in the hundreds place since there are no hundreds, and then a *0* in the tens place since there are no tens, and a *0* in the units place to show there are no units.

13. Write out and have your child read: *702*. If necessary, explain: Since there are no tens in this number, we read *seven hundred two.*

14. Write out and have your child read: *207, 702, 105, 604, 309, 903, 405, 806, 508, 607.*

15. Ask: How many tens in 207? How many units? How many hundreds?

16. Have your child now write on the slate in figures: *four hundred two.* If he needs further direction, you may say: First write the *4* hundreds and then write *02*. We write the *0* to show there are no tens and the *2* to show the two units.

17. Now, going slowly enough to allow for neatness, have him write the following numbers in figures as you dictate: *three hundred six; two hundred five; one hundred eight; six hundred six.*

Addition of Larger Sums without Carrying

Sample *(instructions for the parent)*

Now that your child has worked a bit with money sums using dollars, practice with pure number is given.

Have your child get out his math notebook as you will also show him how to write these sums with the hundreds, tens, and units having their own places. Use a slate or grid paper to demonstrate how it is to be written.

Write the equation on the slate or grid paper.

$$\begin{array}{r} 321 \\ +214 \\ \hline \end{array}$$

Talk through the process of solving it (or allow your student to talk through the process, if he is able) and demonstrate how the answer would be written. (1 unit + 4 units = 5 units, to be written in the units place; 2 tens + 1 ten = 3 tens, to be written in the tens place; 3 hundreds + 2 hundreds = 5 hundreds, to be written in the hundreds place.)

$$\begin{array}{r} 321 \\ +214 \\ \hline 535 \end{array}$$

Sums

Tip: The student may write the full equation on the slate as you speak it, or you may write the equations and have the student write only the answers. Either way, the student should talk through the process of solving until he demonstrates good understanding..

1. $\begin{array}{r} 149 \\ +240 \\ \hline \end{array}$ 2. $\begin{array}{r} 361 \\ +421 \\ \hline \end{array}$ 3. $\begin{array}{r} 511 \\ +254 \\ \hline \end{array}$ 4. $\begin{array}{r} 342 \\ +517 \\ \hline \end{array}$ 5. $\begin{array}{r} 634 \\ +242 \\ \hline \end{array}$

Answers

1. $\begin{array}{r} 149 \\ +240 \\ \hline 389 \end{array}$ 2. $\begin{array}{r} 361 \\ +421 \\ \hline 782 \end{array}$ 3. $\begin{array}{r} 511 \\ +254 \\ \hline 765 \end{array}$ 4. $\begin{array}{r} 342 \\ +517 \\ \hline 859 \end{array}$ 5. $\begin{array}{r} 634 \\ +242 \\ \hline 876 \end{array}$

Addition of Larger Sums with Carrying

Sample *(instructions to the parent)*

Write this equation on grid paper or a slate.

$$\begin{array}{r} 526 \\ +285 \\ \hline \end{array}$$

Now speak it through and demonstrate how to notate each step, involving your child in the process. Begin with the units. (6 units + 5 units is the same as 11 units, or 10 and 1. So that is 1 unit for the units place and what shall we do with our ten? Yes, a ten is to be added to the other tens.)

$$\begin{array}{r} {\scriptstyle 1} \\ 526 \\ +285 \\ \hline 1 \end{array}$$

Proceed to the tens. (2 tens + 8 tens = 10 tens, 10 tens gives us 1 hundred, and there is still 1 ten, making 1 hundred and 1 ten. 1 ten in the tens place and 1 hundred to be added to [the child will most likely be able to tell you at this point] the other hundreds. Or, 1 ten + 2 tens + 8 tens = 11 tens, or 1 hundred and 1 ten.)

$$\begin{array}{r} {\scriptstyle 1\,1} \\ 526 \\ +285 \\ \hline 11 \end{array}$$

Last, solve for the hundreds and give the answer. (1 hundred and 5 hundreds and 2 hundreds give us 8 hundreds. Our answer is 811.)

$$\begin{array}{r} {\scriptstyle 1\,1} \\ 526 \\ +285 \\ \hline 811 \end{array}$$

Sums

Write each equation on a slate or on gridded paper, one at a time. As your student works through each problem aloud, he may choose to write the answer or, as it is solved, you may write it for him, depending on his comfort level with writing.

1.	2.	3.	4.	5.
567	428	246	159	561
+355	+196	+536	+453	+419

Tip: Always encourage your child to add in units, tens *and* hundreds. *If he is carrying over 1* ten *or 1* hundred, *he should name it as such and not merely call it a* 1.

6.	7.	8.	9.	10.
338	557	193	208	186
+ 85	+277	+ 49	+228	+166

Answers

1.	2.	3.	4.	5.
1 1	1 1	1	1 1	1
567	428	246	159	561
+355	+196	+536	+453	+419
922	624	782	612	980

6.	7.	8.	9.	10.
1 1	1 1	1 1	1	1 1
338	557	193	208	186
+ 85	+277	+ 49	+228	+166
423	834	242	436	352

Review

1. Elia wants to gather 14 flowers. She has already gathered 9. How many more does she need?

2. Evienne had 6¢, her sister had 4¢, and her brother had 5¢. How much had they in all?

3. The Smiths planted 3 peach trees, 2 lemon trees, and 3 plum trees. How many trees did they plant altogether?

4. Edward split 9 logs then later split 5 more. How many logs did he split in all?

5. There were 14 cookies on a plate. 8 cookies were eaten. How many remained?

6. What is 8 minus 5?

7. What is 3 added to 3?

8. What is 6 take away 1?

9. What is 4 added to 5?

10. What is 4 added to 6?

Answers

1. 5 flowers
2. 15¢
3. 8 trees
4. 14 logs
5. 6 cookies
6. 3
7. 6
8. 5
9. 9
10. 10

Addition of Longer Sums with Three or More Figures

Tip: Keep lessons to 15 minutes of work on the newest concept and 5 minutes of review at the end, making each day's math time 20 minutes total.

Tip: We never want the act of writing to overtake the idea being presented in arithmetic so you may write each problem on either a slate or gridded paper for your child to solve. Depending on where he is in his comfort with writing, as he works through each problem aloud he may choose to write the answer or, as it is solved, you may write it for him.

Sums

1.	2.	3.	4.	5.
5	6	2	8	7
5	6	7	1	6
+4	+3	+5	+2	+3

Tip: If you need to get the coin bag or stick bundles back out for a time you may do so, always with a view of putting them away as soon as possible.

6.	7.	8.	9.	10.
2	5	6	1	5
7	6	8	9	2
2	3	1	3	5
+9	+4	+6	+7	+3

Answers

1.	2.	3.	4.	5.
1	1	1	1	1
5	6	2	8	7
5	6	7	1	6
+ 4	+ 3	+ 5	+ 2	+ 3
14	15	14	11	16

6.	7.	8.	9.	10.
2	1	2	2	1
2	5	6	1	5
7	6	8	9	2
2	3	1	3	5
+ 9	+ 4	+ 6	+ 7	+ 3
20	18	21	20	15

Tip: Use as many problems as necessary in order to secure understanding and build a comfort with this type of work. Problems not used can be bookmarked and used later for review.

Sums *(cont.)*

11.	12.	13.	14.	15.
3 18 +74	26 32 +36	9 27 +43	14 55 +22	33 26 +53

16.	17.	18.	19.	20.
8 12 31 +32	13 20 49 +11	4 17 25 + 21	16 21 23 +48	15 16 38 +62

Answers

11.	12.	13.	14.	15.
1 3 18 +74 95	1 26 32 +36 94	1 9 27 +43 79	1 14 55 +22 91	1 1 33 26 + 53 112

16.	17.	18.	19.	20.
1 8 12 31 +32 83	1 13 20 49 +11 93	1 4 17 25 +21 67	1 1 16 21 23 + 48 108	1 2 15 16 38 + 62 131

Sums *(cont.)*

21.	22.	23.	24.	25.
52 114 +253	147 134 +262	42 64 +298	52 159 +483	107 251 +342

26.	27.	28.	29.	30.
5 14 85 +291	2 81 317 +421	60 78 115 +694	158 122 214 +363	48 96 233 +432

21.	22.	23.	24.	25.
1	1 1	2 1	1 1	1 1
52	147	42	52	107
114	134	64	159	251
+253	+262	+298	+483	+342
419	543	404	694	700

26.	27.	28.	29.	30.
1 1	1 1	2 1	1 1	2 1
5	2	60	158	48
14	81	78	122	96
85	317	115	214	233
+291	+421	+694	+363	+432
395	821	947	857	809

Review of Addition with Larger Numbers

1. Luca had $1.75 in his piggy bank and then added 24¢ to his piggy bank. How much does Luca now have in his bank?

2. Fiona needs to get some things for her kitten. A collar costs $2.89, catnip costs 75¢, and a toy mouse costs $3.49. What is the total cost?

3. Hank bought a brick set for $4.97 and a case for $2.78. How much total did he spend?

4. How much is 27 added to 312?

Tip: Your child's writing ability and comfort level will dictate how much writing he is to do at this point. If writing takes away from the ideas contained in the lesson, then you may write the problem and have him write the solution or he may solve the exercises orally as you write the solution for him.

5. Add together: 6, 12, and 74.

6. Mary is gathering sewing supplies. Her felt cost 47¢, the thread cost $1.95, and the needle cost $2.44. What was the total cost?

7. Pip buys a package of gum for 80¢, a honey stick for 25¢, and a 53¢ lollipop. How much did he spend in all?

8. Add together 14, 65, and 38.

9. How much are 326 added to 448?

10. A bottle of water costs 65¢ at the stand. Ben buys 4 bottles of water for his family. What was the total that he spent? *(Taken experimentally. This should be written as an addition problem.)*

11. Cole earned $5.15 in the morning and $3.25 in the afternoon. How much did he earn in all?

12. What is 3 + 8 + 11 + 19?

13. What is 25 + 25 + 25?

14. Oleg purchased a bag of popcorn for 50¢, a box of caramel corn for 75¢, and a bag of peanuts for 75¢. How much did he give for all?

15. Hyacinth bought a pink ribbon for 39¢, a red ribbon for 38¢, and a purple ribbon for 42¢. How much did she spend all together?

16. What is 712 added to 190?

17. What does 40 + 40 + 40 equal?

18. At the book sale Molly purchased a book for $2.75, another book for $1.75, and some sheet music for $1.50. How much did Molly spend in all?

19. Roger has 86¢ in his pocket, 55¢ on his desk, and 77¢ in his piggy bank. How much has Roger in all?

20. Add together: 101 + 252 + 368.

Answers

1. $1.99
2. $7.13
3. $7.75
4. 339
5. 92
6. $4.86
7. $1.58
8. 117
9. 774
10. $2.60
11. $8.40
12. 41
13. 75
14. $2.00
15. $1.19
16. 902
17. 120
18. $6.00
19. $2.18
20. 721

Subtraction with Larger Numbers

Now we will continue with subtraction in much the same way as addition was taken.

We begin with a review that reinforces the idea of subtraction as the complement of addition. As the student solves each subtraction equation, he will work it backward using addition to check his answer. We will then proceed to solidify the concept of exchanging dimes and pennies, or tens and units, to introduce what is commonly called "borrowing" or "regrouping" and show how to notate that idea in writing.

As the student progresses, larger numbers into the hundreds will be reviewed from previous lessons and then included in the subtraction problems. Much practice will be given to working with subtraction in larger numbers, both with and without borrowing.

Keep lessons to 15 minutes of work on the newest concept and 5 minutes of review at the end, making each day's math time 20 minutes total. Some natural review is inherent in the lessons, as any subtraction problem can be worked in reverse using addition to check the answer, thus reviewing both operations. You will find additional review sums (for both addition and subtraction) in the back of this book, beginning on page 195, or feel free to reuse sums from previous lessons, changing names, numbers, and objects as desired.

Tip: The term sum *in Charlotte Mason's day referred to any mathematical problem, whether addition or subtraction. It will be used in that context in these printed lessons to the teacher. The modern term* equation *will be used when speaking to the student.*

To prepare for the following lessons, you may wish to watch the demonstrations in the session "Subtraction with Larger Numbers" in Charlotte Mason's *Living Math: A Guided Journey.*

Review Subtraction without Exchanging

Tip: As a help, pull a Number Sentence Card out before the lesson begins so your student's work can continue uninterrupted if you get called away for a moment.

Review Sums

Objects used: coins

1. If I have 3 books but want 5, how many more must I get?
2. If you picked 2 apples but need to pick 6, how many more must you pick?
3. If you have 6¢ but need 9¢, how much more must you get?
4. If you have 8¢ and give me 1¢, how much have you left?
5. If you had 9¢ in your pocket and lost 2¢, how much have you left?
6. I had 8¢ but I lost 3¢. How much have I left?
7. I had 7¢ and spent 5¢. How much remains?
8. Take away 2¢ from 10¢ and how much remains?
9. There are 9¢ on the table and you take away 4¢. How much remains?
10. There are 15¢ on the table and you take away 11¢. How much remains?

Answers

1. 2 books
2. 4 apples

3. 3¢
4. 7¢
5. 7¢
6. 5¢
7. 2¢
8. 8¢
9. 5¢
10. 4¢

Sample *(instructions to the parent)*

Objects used: coins

Now your child will subtract and then use his answer to add the equation backward and check his work.

Tell your child: Take 58¢ (as 5 dimes and 8 pennies) out of your coin bag. Take away 12¢. What have you left? Have your student speak this process through (something like, 8 pennies take away 2 pennies leave 6 pennies and 5 dimes take away 1 dime leave 4 dimes).

○●

$$\begin{array}{r} 58 \\ -12 \\ \hline 46 \end{array}$$

Tip: The student should work only with the coins on the table at this point. Written notation will be included later. The written sums in this section are just for the teacher's benefit. The white dot indicates the dimes column; the black dot indicates the pennies column.

Now, keeping the above sum all out on the table, lead the child to add this back again (from the bottom to the top) to reinforce the idea of subtraction as the complement of addition (such as, 6 pennies and 2 pennies make 8 pennies, and 4 dimes and 1 dime make 5 dimes).

Sums

1. Begin with 77¢. Take away 24¢ (or 2 dimes and 4 pennies from 7

dimes and 7 pennies). What is left? (4 pennies taken from 7 pennies leaves 3 pennies and 2 dimes taken from 7 dimes leaves 5 dimes.)

Now add it back up again (3 pennies and 4 pennies make 7 pennies and 5 dimes and 2 dimes make 7 dimes).

○●

$$\begin{array}{r} 77 \\ -24 \\ \hline 53 \end{array}$$

2. Begin with 98¢. Take away 3 dimes and 4 pennies from 9 dimes and 8 pennies. What is left? (4 pennies taken from 8 pennies leaves 4 pennies and 3 dimes taken from 9 dimes leaves 6 dimes.)

   And add it back up (4 pennies and 4 pennies make 8 pennies and 6 dimes and 3 dimes make 9 dimes).

○●

$$\begin{array}{r} 98 \\ -34 \\ \hline 64 \end{array}$$

3. Begin with 49¢. Take away 1 dime and 7 pennies from 4 dimes and 9 pennies. What remains? (7 pennies taken from 9 pennies leaves 2 pennies and 1 dime taken from 4 dimes leaves 3 dimes.)

   Add it back up (2 pennies and 7 pennies make 9 pennies and 3 dimes and 1 dime make 4 dimes).

○●

$$\begin{array}{r} 49 \\ -17 \\ \hline 32 \end{array}$$

4. Begin with 57¢. Take away 3 dimes and 3 pennies from 5 dimes and 7 pennies. What is left? (3 pennies taken from 7 pennies leaves 4 pennies and 3 dimes taken from 5 dimes leaves 2 dimes.)

Add it back up (4 pennies and 3 pennies make 7 pennies and 2 dimes and 3 dimes make 5 dimes).

○●
$$\begin{array}{r} 57 \\ -33 \\ \hline 24 \end{array}$$

Review Subtraction with Exchanging

Tip: Keep lessons to 15 minutes of work on the newest concept and 5 minutes of review at the end, making each day's math time 20 minutes total.

Sample *(instructions to the parent)*

Objects used: coins

Once your child has a solid grasp of subtraction as outlined above, we introduce "borrowing" in the concrete, still using dimes and pennies.

Have your child take 2 dimes out of his coin bag. Ask: You have 20¢ and you give a friend 8¢. What have you left?

It will probably occur to him after a moment to change one of his dimes into pennies. Once he takes the 8 pennies away, he will show that 20¢ take away 8¢ leaves 12¢, or 1 dime and 2 pennies.

Sums

1. You have 10¢ and you spend 4¢. What do you have left?
2. You have 20¢ and you give someone 7¢. What remains?
3. You have 15¢ and you give away 10¢. What do you have left?

4. You had 37¢ and spent 18¢. What did you have left?

5. You have 28¢ and then give a friend 9¢. What is left?

6. You earned 54¢ and spent 46¢. How much do you still have?

7. You saved 86¢, then gave 48¢ away. What do you have left?

8. You had 54¢ and you bought a stuffed animal for 45¢. How much do you have left?

9. You earned 98¢ and you gave your friend 49¢ for helping you. How much remains for you?

10. You have 20¢ and you gave a friend 8¢. What have you left?

Answers

1. 6¢
2. 13¢
3. 5¢
4. 19¢
5. 19¢
6. 8¢
7. 38¢
8. 9¢
9. 49¢
10. 12¢

Subtraction Practice with Notation

Samples *(instructions to the parent)*

Objects used: coins

1. Tell your child: Take out 23¢ in dimes and pennies. Now pay me 6¢

from what you have. How much have you left? Your child will change a dime into ten pennies in order to "pay" you. (Answer: 23 – 6 = 17)

After the child has given you the answer, have him write the sum in his math notebook in a column if he already knows how. If not, you will show him first on the slate, speaking something like the following as you write:

23¢ take away 6¢

$$\begin{array}{r} 23 \\ -\ \ 6 \\ \hline \end{array}$$

We couldn't take 6 units from our 3 so we changed one of our tens into units, giving us 1 ten and 13 units.

$$\begin{array}{r} {}^{1}\cancel{2}{}^{1}3 \\ -\ \ 6 \\ \hline \end{array}$$

So 6 units taken from 13 units leaves 7 units, and nothing taken from 1 ten leaves 1 ten. 23 – 6 = 17.

$$\begin{array}{r} {}^{1}\cancel{2}{}^{1}3 \\ -\ \ 6 \\ \hline 17 \end{array}$$

2. Have your child work this equation: Now take 40¢ out in dimes. You have 40¢ and you need to take away 24¢.

Once the child has worked this problem in the concrete, changing dimes to pennies to get his answer, have him write the entire sum. He may do so first on his slate with your help if needed, talking through each step with you, and then copy it into his notebook.

$$\begin{array}{r} {}^{3}\cancel{4}{}^{1}0 \\ -24 \\ \hline 16 \end{array}$$

Sums

Now give another few problems, this time to work solely in the math notebook, seeing if he is able to mentally change the dimes into pennies.

1. You have 50¢ and you take away 25¢. What have you left?

Tip: Occasionally have your student check his answer by working the problem in reverse using addition.

2. You have 21¢ and you take away 5¢. What have you left?
3. You have 41¢ and you take away 36¢. What have you left?
4. You have 90¢ and you take away 15¢. What have you left?
5. You have 32¢ and you take away 29¢. What have you left?
6. You have 75¢ and you take away 38¢. What have you left?
7. You have 17¢ and you take away 8¢. What have you left?
8. You have 80¢ and you take away 44¢. What have you left?

Answers

1. 25¢
2. 16¢
3. 5¢
4. 75¢
5. 3¢
6. 37¢
7. 9¢
8. 36¢

Subtraction in Pure Number with Notation

Sample *(instructions to the parent)*

Object used: craft sticks (some loose and some in ten bundles) if necessary

Since your child has become accustomed to changing dimes into pennies, it should be simple for him to change tens into units.

Take out the dry erase board or slate. Tell your child: Now we are going to subtract or take away 24 from 31. Do you know how it is written? Pause and give your child time to show you if he knows. If not, explain as outlined below.

It is written like so, keeping the units in their place and the tens in their place:

$$\begin{array}{r} 31 \\ -24 \\ \hline \end{array}$$

We cannot take 4 units from 1 unit so what should we do?

At this point, your child will probably suggest turning a ten into units. If not, take out your craft stick bundles and have your child solve the problem by breaking up a ten bundle. When he has solved it, have him write the equation in his math notebook.

$$\begin{array}{r} \overset{2}{\cancel{3}}{}^{1}1 \\ -\ 24 \\ \hline 7 \end{array}$$

Sums

Tip: Be sure to work at your child's pace and ensure he is writing neatly in his notebook with special care for place value.

1. 40	2. 63	3. 97	4. 71	5. 54
−15	−44	−39	−48	−17

Tip: Occasionally have your student check his answer by working the problem in reverse using addition.

Answers

1. ~~4~~^{3}^{1}0	2. ~~6~~^{5}^{1}3	3. ~~9~~^{8}^{1}7	4. ~~7~~^{6}^{1}1	5. ~~5~~^{4}^{1}4
−15	−44	−39	−48	−17
25	19	58	23	37

Subtraction Using Dollars without Borrowing

Tip: Keep lessons to 15 minutes of work on the newest concept and 5 minutes of review at the end, making each day's math time 20 minutes total.

Samples *(instructions to the parent)*

Objects used: dollars, dimes, pennies

Your next step will be the same as in addition, but this time you will lead the child in subtraction of larger numbers using dollars. You can use real money or play money. Work a few equations in the concrete, taking care to have the money in place-value order on the table: dollars to the student's left, dimes to the right of the dollars, and then pennies to the right of dimes.

1. Ask your child to count out \$9.84 in dollars, dimes, and pennies. Now have your child take away or subtract \$5.61 from that amount, working with the concrete objects.

Tip: Have your child speak through his steps, e.g., 1 penny from four pennies is 3 pennies. 6 dimes from 8 dimes leaves 2 dimes. 5 dollars from 9 dollars leaves 4 dollars.

$$\begin{array}{r} \$9.84 \\ -5.61 \\ \hline \$4.23 \end{array}$$

2. Have your child work through a couple more such problems, speaking through them as outlined in the example above.

$$\begin{array}{r} \$3.75 \\ -2.51 \\ \hline \$1.24 \end{array} \qquad \begin{array}{r} \$7.48 \\ -4.13 \\ \hline \$3.35 \end{array}$$

3. Now you will show how to write the next problem as he solves it. Saying something like: Let's write out this next problem as you solve it. It is written like so, keeping the hundreds, tens, and units all in their places.

   (This example details the process step by step, but you will write the equation only once on the slate for your student, then add each step as he solves it).

take away units $\begin{array}{r} \$5.62 \\ -3.21 \\ \hline 1 \end{array}$ *then dimes* $\begin{array}{r} \$5.62 \\ -3.21 \\ \hline 41 \end{array}$ *then dollars* $\begin{array}{r} \$5.62 \\ -3.21 \\ \hline \$2.41 \end{array}$

Sums

Tip: Have your child speak through each problem as he works it until he demonstrates good understanding. You may have him write a problem from time to time.

Tip: Be content to go slowly as your child develops understanding, concentration, and accuracy. Use as many problems as necessary for your child to gain confidence in working this type of problem. Don't move on until your child has truly grasped each idea presented and feels comfortable working with it.

1. $3.47 –1.26	2. $4.84 –2.63	3. $8.97 –5.64	4. $4.27 –2.14	5. $5.73 –4.62

Tip: Occasionally have your student check his answer by working the problem in reverse using addition.

6. $2.47 –1.35	7. $9.75 –4.65	8. $7.44 –2.31	9. $5.68 –4.25	10. $7.97 –3.25
11. $5.68 –2.44	12. $2.57 –1.36	13. $9.65 –7.44	14. $4.78 –2.65	15. $8.47 –2.23
16. $4.54 –3.41	17. $2.37 –2.10	18. $8.75 –5.23	19. $9.48 –3.25	20. $4.92 –3.40

Answers

1. $3.47 –1.26 $2.21	2. $4.84 –2.63 $2.21	3. $8.97 –5.64 $3.33	4. $4.27 –2.14 $2.13	5. $5.73 –4.62 $1.11
6. $2.47 –1.35 $1.12	7. $9.75 –4.65 $5.10	8. $7.44 –2.31 $5.13	9. $5.68 –4.25 $1.43	10. $7.97 –3.25 $4.72
11. $5.68 –2.44 $3.24	12. $2.57 –1.36 $1.21	13. $9.65 –7.44 $2.21	14. $4.78 –2.65 $2.13	15. $8.47 –2.23 $6.24
16. $4.54 –3.41 $1.13	17. $2.37 –2.10 $ .27	18. $8.75 –5.23 $3.52	19. $9.48 –3.25 $6.23	20. $4.92 –3.40 $1.52

Subtraction Using Dollars with Borrowing

Sample *(instructions to the parent)*

Objects used: dollars, dimes, pennies

Give your child the sum below that requires changing a dollar into dimes. Have your child explain the process as he works through it (e.g., I've 0 cents to take from 4 cents, so I still have 4 cents. I cannot take 5 dimes from 3 dimes so must change 1 of my dollars into dimes. 5 dimes taken from 13 dimes is 8 dimes. 1 dollar from 3 dollars leaves 2 dollars).

Tip: This step should be relatively simple for him since he is used to changing dimes to pennies and tens to units.

$$\begin{array}{r} \$4.34 \\ \underline{-1.50} \\ \$2.84 \end{array}$$

Once the answer is obtained, have your student write the whole sum in his math notebook. At this point it is probably already evident how it will be written but, if not, take him through step-by-step on the slate first.

$$\begin{array}{r} {}^{3}\not{4}.{}^{1}34 \\ \underline{-\ 1.50} \\ 2.84 \end{array}$$

Sums

Tip: Have your child speak through each problem as he works it and give him practice writing an occasional equation.

1. $\begin{array}{r} \$3.48 \\ \underline{-2.57} \end{array}$ 2. $\begin{array}{r} \$4.86 \\ \underline{-3.95} \end{array}$ 3. $\begin{array}{r} \$7.17 \\ \underline{-2.93} \end{array}$ 4. $\begin{array}{r} \$4.33 \\ \underline{-1.71} \end{array}$ 5. $\begin{array}{r} \$5.84 \\ \underline{-3.68} \end{array}$

Tip: Occasionally have your student check his answer by working the problem in reverse using addition.

6. $8.50 −2.79	7. $6.44 −3.84	8. $5.86 −3.93	9. $7.36 −4.69	10. $9.15 −4.86
11. $9.10 −4.86	12. $5.10 −4.40	13. $8.49 −3.58	14. $3.47 −2.56	15. $3.82 −1.49
16. $6.54 −4.56	17. $9.86 −5.48	18. $4.57 −3.92	19. $7.35 −2.84	20. $8.37 −4.84

Tip: Use as many problems as are needed until your student is working with ease. Once you begin the next concept you may use any remaining problems in this section for review work.

Answers

2 1 1. $~~3~~.48 −2.57 $ .91	3 1 2. $~~4~~.86 −3.95 $ .91	6 1 3. $~~7~~.17 −2.93 $4.24	3 1 4. $~~4~~.33 −1.71 $2.62	7 1 5. $5.~~8~~4 −3.68 $2.16
7 14 1 6. $~~8~~.~~5~~0 −2.79 $5.71	5 1 7. $~~6~~.44 −3.84 $2.60	4 1 8. $~~5~~.86 −3.93 $1.93	6 12 1 9. $~~7~~.~~3~~6 −4.69 $2.67	8 10 1 10. $~~9~~.~~1~~5 −4.86 $4.29
8 10 1 11. $~~9~~.~~1~~0 −4.86 $4.24	4 1 12. $~~5~~.10 −4.40 $ .70	7 1 13. $~~8~~.49 −3.58 $4.91	2 1 14. $~~3~~.47 −2.56 $ .91	7 1 15. $3.~~8~~2 −1.49 $2.33
5 14 1 16. $~~6~~.~~5~~4 −4.56 $1.98	7 1 17. $9.~~8~~6 −5.48 $4.38	3 1 18. $~~4~~.57 −3.92 $ .65	6 1 19. $~~7~~.35 −2.84 $4.51	7 1 20. $~~8~~.37 −4.84 $3.53

Subtraction in Pure Number with Borrowing

Tip: As a help, pull a Number Sentence Card out before the lesson begins so your student's work can continue uninterrupted if you get called away for a moment.

Tip: You may begin by writing one of the problems on a slate and having your child speak through the steps as he solves it until he is mentally able to change tens to units and hundreds to tens. The amount of writing a child is able to accomplish at this time varies greatly. You never want the writing to overtake the significance of the ideas in the lesson. You might begin with your student copying a problem every so often into his math notebook after he has solved it on the slate.

Tip: If your child needs to use dollars, dimes, and pennies to do the first few problems that is fine, but try to work with a view to putting the objects away.

1. $\begin{array}{r} 145 \\ -\ 32 \\ \hline \end{array}$ 2. $\begin{array}{r} 236 \\ -\ 83 \\ \hline \end{array}$ 3. $\begin{array}{r} 717 \\ -265 \\ \hline \end{array}$ 4. $\begin{array}{r} 565 \\ -239 \\ \hline \end{array}$ 5. $\begin{array}{r} 383 \\ -191 \\ \hline \end{array}$

Tip: Occasionally have your student check his answer by working the problem in reverse using addition.

6. $\begin{array}{r} 586 \\ -247 \\ \hline \end{array}$ 7. $\begin{array}{r} 984 \\ -391 \\ \hline \end{array}$ 8. $\begin{array}{r} 655 \\ -195 \\ \hline \end{array}$ 9. $\begin{array}{r} 275 \\ -186 \\ \hline \end{array}$ 10. $\begin{array}{r} 494 \\ -385 \\ \hline \end{array}$

Answers

1. $\begin{array}{r} 145 \\ -\ 32 \\ \hline 113 \end{array}$ 2. $\begin{array}{r} \overset{1}{\not{2}}\,{}^{1}36 \\ -\ 83 \\ \hline 153 \end{array}$ 3. $\begin{array}{r} \overset{6}{\not{7}}\,{}^{1}17 \\ -265 \\ \hline 452 \end{array}$ 4. $\begin{array}{r} 5\overset{5}{\not{6}}\,{}^{1}5 \\ -239 \\ \hline 326 \end{array}$ 5. $\begin{array}{r} \overset{2}{\not{3}}\,{}^{1}83 \\ -191 \\ \hline 192 \end{array}$

6. $\begin{array}{r} 5\overset{7}{\not{8}}\,{}^{1}6 \\ -247 \\ \hline 339 \end{array}$ 7. $\begin{array}{r} \overset{8}{\not{9}}\,{}^{1}84 \\ -391 \\ \hline 593 \end{array}$ 8. $\begin{array}{r} \overset{5}{\not{6}}\,{}^{1}55 \\ -195 \\ \hline 460 \end{array}$ 9. $\begin{array}{r} \overset{1}{\not{2}}\,\overset{16}{\not{7}}\,{}^{1}5 \\ -186 \\ \hline 89 \end{array}$ 10. $\begin{array}{r} 4\overset{8}{\not{9}}\,{}^{1}4 \\ -385 \\ \hline 109 \end{array}$

Review of Subtraction with Larger Numbers

Tip: If your child is able to write each equation on his slate while you give the problem, he may do so, but be watchful that the act of writing does not become more important than the ideas being presented. A problem here and there may be copied neatly into the math notebook after solving.

Tip: Always be content to go at your child's pace.

1. Farmer Jack has 250 apple trees and 175 pear trees in his orchard. How many more apple trees has he than pear trees? Find the difference:

$$\begin{array}{r} {}^{1\;14\;1} \\ \cancel{2}\cancel{5}0 \\ -\,175 \\ \hline 75 \end{array}$$

2. Ren picked 32 apples and Jill picked 17. How many more apples did Ren pick than Jill?

3. On their day at the beach, Milli collected 37 shells, Molly collected 41, and May collected 28. How many shells were collected in all?

4. Since Molly collected 41 shells and May collected 28, how many more shells did Molly collect than May?

5. Mrs. Downing is collecting buttons for a garland she wants to make. She has collected 78 but needs 135. How many more buttons must she collect?

6. Luke has 43 toy cars and Jack has 57. How many more cars does Jack have than Luke?

7. Sean has $4.87. He spent $2.64 on a dog toy. How much money has he left?

8. Gage gathers 16 pieces of paper for his craft project. He uses 8 pieces. How many does he have to put away?

9. Wyatt earned $7.94 and bought a $5.86 board game to play with Abby. How much money does he have left?

10. Sarina had 39 straws and distributed 27 at the picnic. How many does she still have?

11. Gabe was given $2.44 for helping his mother and bought a 39-cent balloon. How much money does he have left?

12. Thomas is putting together a 64-piece puzzle. So far he has used 59 pieces. How many puzzle pieces remain to complete the puzzle?

13. New watercolor paints cost $4.38 and Maddy has $2.76. How much more does she need to purchase the paints?

14. Abby is selling lemonade and cookies at her stand. She earned a total of $5.89, but owes her mom $1.48 for all the ingredients. What are her final earnings?

15. Fiona helped her mom make 14 waffles and they ate 5. How many waffles remain on the platter?

16. $\begin{array}{r} 9 \\ -5 \\ \hline \end{array}$ 17. $\begin{array}{r} 23 \\ -11 \\ \hline \end{array}$ 18. $\begin{array}{r} 45 \\ -36 \\ \hline \end{array}$ 19. $\begin{array}{r} 681 \\ -\ 59 \\ \hline \end{array}$ 20. $\begin{array}{r} 974 \\ -484 \\ \hline \end{array}$

Answers

1. 75 apple trees
2. 15 apples
3. 106 shells
4. 13 shells
5. 57 buttons
6. 14 cars
7. $2.23

8. 8 pieces
9. $2.08
10. 12 straws
11. $2.05
12. 5 pieces
13. $1.62
14. $4.41
15. 9 waffles

16.
$$\begin{array}{r} 9 \\ -5 \\ \hline 4 \end{array}$$

17.
$$\begin{array}{r} 23 \\ -11 \\ \hline 12 \end{array}$$

18.
$$\begin{array}{r} \overset{3}{\cancel{4}}\,{}^{1}5 \\ -36 \\ \hline 9 \end{array}$$

19.
$$\begin{array}{r} 6\overset{7}{\cancel{8}}\,{}^{1}1 \\ -\ \ 59 \\ \hline 622 \end{array}$$

20.
$$\begin{array}{r} \overset{8}{\cancel{9}}\,{}^{1}74 \\ -484 \\ \hline 490 \end{array}$$

Formal Introduction of Multiplication

We are at the point in our journey when multiplication is formally introduced. Your child will have already experienced a natural overlap with multiplication, which occurs during the investigation of numbers and addition. He will have seen that "two 2s make 4," having laid out rows of twos with beans, and "three 2s make 6" in Book 1. He will also have added in 2s or 3s, etc.

Once your child has an idea of what "times" indicates, written tables are begun. Multiplication tables are learned by heart in a Charlotte Mason classroom but care is taken to prepare for the task in the best way possible. A child feels a comfort with numbers that he's had plenty of time to investigate and explore. He's already proven many facts in the concrete and been introduced to certain ideas, such as the relationship between addition and multiplication, and that these facts will always work, whether counting apples or dollars.

We will begin by introducing the idea of multiplication as repeated addition, guiding the child to discover what is meant by the term "times." Multiplication tables will then be built with concrete objects and plenty of work given on each table to help students thoroughly understand and memorize it. The work in this book will cover multiplication tables for 1 through 6; tables 7 through 10 will be covered in Book 3. After each table is learned, that number will subsequently be used as the multiplier with a larger number. Some multiplication using a double-digit multiplier will also be given experimentally.

Lessons should continue to last no longer than 15 minutes, with 5 more minutes of review at the end, for a total of 20 minutes. Review can include addition, subtraction, and multiplication covered so far. You will find review sections in the lessons as well as more review sums in the back of this book, beginning on page 296. Also remember that you can use sums from any previous lessons as part of your review time.

You may wish to view the sessions on "Multiplication," "Constructing Multiplication Tables," and "Multiplication of Larger Numbers" in *Charlotte Mason's Living Math: A Guided Journey* for more detail and demonstrations.

Multiplication as Repeated Addition

You will need a slate or gridded paper for yourself as well as a slate for each child. Your child will also be using his coin bag full of pennies and dimes. To begin, we will reinforce multiplication as repeated addition through simple, interesting problems using coins and other concrete objects.

Introduction of "Times" *(instructions to the parent)*

You will be writing on the slate until you instruct your child to do so. We begin with very simple questions where knowing the multiplication table isn't necessary. Here it is broken down into steps for you:

1. Say: John had 2¢ and a friend gave him 2¢ more. How many cents had he then? (4¢)
   Write on the board: 2 + 2 = 4
   Now ask: How many times do we have 2 cents? (2 times)

2. Ask: If 3 children had 2¢ each, how much had they all together? (6¢)
   Write on the board: 2 + 2 + 2 = 6
   Ask: How many times do we have 2 cents? (3 times)

3. Say: You bought five gum balls at 2¢ apiece. How much did they cost all together? (10¢)
   Write on the board: 2 + 2 + 2 + 2 + 2 = 10
   Ask: How many times do we have 2 cents? (5 times)

4. Say: Macy bought 4 candies at 4¢ apiece. How much did they cost all together? (16¢)
   Write on the board: 4 + 4 + 4 + 4 = 16
   Ask: How many times do we have 4 cents? (4 times)

5. Say: You bought five apples at 7¢ apiece. How much did they cost all together? (35¢)
   Write on the board: 7 + 7 + 7 + 7 + 7 = 35

Ask: How many times do we have 7 cents? (5 times)

6. Say: Jack saved 5¢ every day for 3 days. How much did he save in total? (15¢)
   Write on the board: 5 + 5 + 5 = 15
   Ask: How many times do we have 5 cents? (3 times)

Introduction of the "x" Symbol *(instructions to the parent)*

Now state: We can write this in a shorter way. Let's take a look at a problem we've done before:

1. Say: You bought five gum balls at 2¢ apiece. How much did they cost all together? (10¢ or 1 dime)
2. Write on the board: 2 + 2 + 2 + 2 + 2 = 10
3. Ask: How many times do we have 2 cents? (5 times)
4. Explain: We can write it this way: 2 × 5 = 10
5. Point to (or circle) the "× 5." Tell your child that it means multiplied by 5 —that is, each of the quantities is to be taken 5 times, so that 2 × 5 means five twos.
6. Explain: The sign "×" is read *multiplied by* and means the first is to be multiplied by the second; so "2 × 5 = 10" shows that 2 multiplied by 5 = 10.
7. Also explain: The " ×" sign may also be read *times*. 2 x 5 shows 2 is taken five times.
   So 4 × 3 would mean 4 *multiplied by* 3 = 12, or 4 *times* 3 or 4 taken 3 times.

Sums

Now have your child work a few problems on his slate, using the multiplication sign, then have him read the entire multiplication sentence.

1. If 4 children had 5¢ each, how much had they all together?

2. Gum balls cost 2¢. How much must you pay for 6 gum balls?

3. 3 girls have 3 ribbons each. How many ribbons have they all together?

4. 4 children collected 4 feathers each. How many feathers do they have together?

5. A florist puts 6 flowers in every bouquet. How many total flowers are in 3 bouquets?

6. 4 nests each held 3 eggs. How many total eggs are there?

Answers

Tip: Your child may word these in different ways. As long as the wording and answer are correct, it does not have to specifically match the wording below.

1. $5 \times 4 = 20$; i.e., 5¢ taken 4 times equals 20¢
2. $2 \times 6 = 12$; i.e., 2¢ times 6 equals 12¢
3. $3 \times 3 = 9$; i.e., 3 multiplied by 3 equals 9 ribbons
4. $4 \times 4 = 16$; i.e., 4 times 4 equals 16 feathers
5. $6 \times 3 = 18$; i.e., 6 multiplied by 3 equals 18 flowers
6. $4 \times 3 = 12$; i.e., 4 times 3 equals 12 eggs

Multiplication Table for 2

Tip: Since we are working with the idea of what times *indicates, any number taken just one time, or once, remains the same. This idea is found in each table, both constructed and written, as well as in the subsequent table work.*

Tip: Keep lessons to 15 minutes of work on the newest concept and 5 minutes of review at the end, making each day's math time 20 minutes total.

Step 1: Construct the table. *(instructions to the parent)*

Objects used: coins

To solidify understanding, you and your child will make a simple multiplication table using concrete objects in order to see its rationale before creating written tables.

Tip: It is up to you whether to include this concrete activity as future tables are learned. Some students may need to see the concept only once; others may benefit from repeating the activity for each table studied.

Say: Let's take our coins and make a multiplication table for 2. We'll make 10 rows of coins with 2 coins in each row. Let your student read down the column of coins and answer as you guide him:

••	
••	*2 and 2 are (4),*
••	*and 2 are (6),*
••	*and 2 are (8),*
••	*and 2 are (10),*
••	*and 2 are (12),*
••	*and 2 are (14),*
••	*and 2 are (16),*
••	*and 2 are (18),*
••	*and 2 are (20).*

Ask questions that cover the lines of the table out of order.

1. How many 2s are in 10? (5) Remark: So it is right to say 2 x 5 = 10.

2. If a toy costs 2¢ each, how much would 6 toys cost? (12¢)

3. 2 x 4 = (8)

4. 2 x 3 = (6)

5. 2 x 7 = (14)

6. How many 2s in 14? (7)

7. How many 2s in 6? (3)

8. How many 2s in 18? (9)

9. 2 taken 8 times? (16)

10. 2 taken 2 times? (4)

Now you are going to help your child construct a written multiplication table. The way we do this will act as a tool to understanding the rationale behind the table. Again, you will want your slate and gridded math notebook. These will be used in the horizontal position. We will be taking each number here up to x 10.

Tip: In this book, Book 2, the child will be learning tables up through the 6 table; the tables for 7 through 10 will be presented in Book 3.

Here is a finished multiplication table for 2. A multiplication table in a Charlotte Mason math lesson is formatted like several vertical equations in a row, rather than a grid such as you may have learned in elementary school. **Notice that the multiplier is written smaller than the multiplicand, in order to focus attention on the main number of the particular table.** Follow the detailed instructions below to walk your child through creating this table little by little.

1	2	3	4	5	6	7	8	9	10
2	2	2	2	2	2	2	2	2	2
2	4	6	8	10	12	14	16	18	20

Write down the numeral 2 on the (horizontal) slate or graph paper leaving room both above and below:

2

Ask: How many 2s have we? Child answers: 1

Write a small 1 above the 2:

1
2

Ask: One 2 is how many? Child answers: 2

Write a 2 below:

1
2
2

Next write another 2 beside the first one. Point to the middle line and ask, How many 2s have we? Child answers: Two 2s. Write a small 2 above the second 2:

1	2
2	2
2	

Ask: How much is two 2s? Child answers: 4

Write 4 underneath the second 2:

1	2
2	2
2	4

Now add a third 2 to the middle line:

1	2	
2	2	2
2	4	

Say something like, Having three 2s, or 2 "three times," we now have how many all together?

Child answers: 6

Tip: If, at any time, this is not clear, get out objects such as pennies or beans to show it in the concrete.

1	2	3
2	2	2
2	4	6

Continue in the same way until the whole table is written.

Tip: The multiplication tables in this book will go through 10 x the number. If you prefer, feel free to extend it through 12 x the number.

1	2	3	4	5	6	7	8	9	10
2	2	2	2	2	2	2	2	2	2
2	4	6	8	10	12	14	16	18	20

Step 2: Say it through and write it. *(instructions to the parent)*

Once the entire multiplication table for 2 is constructed, give your child a few minutes to look at it and try to visualize it in his head; then have him say it through several times: One 2 is 2, two 2s are 4, three 2s are 6, four 2s are 8, etc.

Erase a few numbers and ask your child to fill them in. For example:

1	2	3	4		6	7		9	10
2	2	2	2	2	2	2	2	2	2
2	4		8	10		14	16		20

Have your child say the table through again. Now erase different numbers and allow him to fill the table in again, then have him repeat the table aloud once more. For example:

1		3	4	5			8		10
	2	2	2	2	2	2	2	2	2
2		6		10	12	14		18	

Tip: The partial tables given here are only samples. Feel free to erase whichever numbers, and how many numbers, you prefer each time you repeat the process.

Have the child write the table in his gridded math notebook, turned in a horizontal position, for ease in referral and as an aid in committing it to memory. As he writes his table, be sure he pays attention to neatness and keeping the proper place value with one number per grid square.

1	2	3	4	5	6	7	8	9	10
2	2	2	2	2	2	2	2	2	2
2	4	6	8	10	12	14	16	18	20

Step 3: Practice with Table Work until learned.

The next step is to give your child a variety of questions. Give only as many as you are able in the time allotted. Questions may be spread out over a few days. He may refer to his written table until it is learned by heart.

1. Six 2s are?
2. How much is 2 taken 7 times?
3. 2 taken 5 times is?
4. How many 2s in 12?
5. How many 2s in 16?
6. How much do 7 marbles cost at 2¢ each?
7. Each hen had 2 chicks. If I have 3 hens, how many chicks are there?
8. Find the cost of 2 pencils at 9¢ each.
9. How many $2 pies am I able to buy with $12?
10. How many 2s make 8?
11. 2 x 4 is?
12. 2 x 9 is?
13. How many 2s in 14?

14. How many 2s in 18?

15. How many 2s in 6?

16. How many $2 cards can you buy with $6?

17. 2 x 10 is?

18. How many 2s in 10?

19. 2 taken 2 times is?

20. If 4 children each had 2 balloons, how many balloons in all?

21. If 8 children each had 2 pencils, how many pencils all together?

22. 2 x 3 is?

23. 2 x 8 is?

24. How many 2s in 4?

25. 2 taken 1 time is?

Answers

1. 12
2. 14
3. 10
4. 6
5. 8
6. 14¢
7. 6 chicks
8. 18¢
9. 6 pies
10. 4
11. 8

12. 18
13. 7
14. 9
15. 3
16. 3 cards
17. 20
18. 5
19. 4
20. 8 balloons
21. 16 pencils
22. 6
23. 16
24. 2
25. 2

Step 4: Work with larger multiplication by 2.

(instructions to the parent)

Objects used: dimes and pennies

Now that your child understands the idea of multiplication and what "times" indicates, he will begin some work in multiplying larger numbers. Work is begun in the concrete, using a majority of money problems, as you explain how the problem is then to be written, while also giving proper terminology in the initial lessons.

1. With the coin bag out, ask your child to take out 34¢ in dimes and pennies, being sure to place dimes on his left and pennies on his right to reinforce the idea of place value.

2. Say: You want to purchase 2 pencils at 34¢ each. How much would you pay in all?

3. The child should get out another 34¢ in dimes and pennies to place under the initial 34¢ in order to tell you that he would pay 68¢.

4. Taking out the slate, write:

$$\begin{array}{r} 34 \\ +34 \\ \hline 68 \end{array}$$

and say: We could find the answer by adding 34¢ + 34¢ or a shorter way would be to multiply 34 by 2, or take 34 times 2.

5. Show how this will be written:

$$\begin{array}{r} 34 \\ \text{x} \quad 2 \\ \hline \end{array}$$

6. Then see if your child can tell you how the equation would be worked and where each number should be written based on the dimes and pennies he has just used; for example, "taking the units first, 4 taken twice (or two 4s, 4 times 2, etc.) with the 8 in the units place."

$$\begin{array}{r} 34 \\ \text{x} \quad 2 \\ \hline 8 \end{array}$$

Then "3 tens taken twice, or 3 tens times 2, with the 6 in the tens place."

$$\begin{array}{r} 34 \\ \text{x} \quad 2 \\ \hline 68 \end{array}$$

7. Now explain the terminology.

34	Multiplicand—the number to be taken; this number does not change and can be taken any number of times; always a *concrete* number.
x 2	Multiplier—how many times the multiplicand will be taken.
68	Product—the product is our result.

Tip: As the numbers grow larger, be sure that the amount of writing done is not too much for your child. Attention to the problem at hand is more important than the effort to write. You may often write the problem, letting him solve it and telling you where each number is placed.

Repeat the steps with a few more shopping questions. Be sure to have your child exchange pennies for dimes as needed. He may write a problem or two out on his slate after solving in the concrete. Also, ask every so often for him to point out the multiplicand, multiplier, or product in any order.

1. Lollipops cost 24¢ apiece. How much money is needed for 2 lollipops?

2. Honey sticks cost 18¢ each. How much for 2 honey sticks?

Tip: Based on the child's previous work in addition and the act of changing pennies to dimes and dimes to dollars, your child will most likely see on his own to begin with units and "carry" over any tens without your having to state the rule for him. If not, simply have him use pennies, dimes, and dollars when working the problem in order to see what is taking place in the concrete. Always be sure your child is multiplying in tens *and* units.

3. Bubblegum is 30¢ apiece. How much should you give for 2 pieces?

4. Erasers cost 42¢. If you buy 2, how much should you give the cashier?

5. Miniature figurines are 27¢. How much for 2? *To be copied neatly in the math notebook after solving.*

6. Ribbons are 33¢ each. How much would 2 ribbons cost?

Tip: Your child may be able to double these easily. That is okay and will reinforce the idea of the relationship between addition and multiplication while giving him confidence. Remember to have him tell you which number is the multiplicand, the multiplier, and the product. As the multiplier grows larger in the following lessons, your student should begin to solve by using multiplication rather than addition.

7. At 17¢ each for a sheet of stickers, how much would 2 sticker sheets be?

8. If a rubber ball costs 26¢, how much for 2?

9. A sparkler costs 35¢ each. How much would you pay for 2 sparklers?

10. An ice cream is 45¢. How much should you give for 2 ice creams? *To be copied neatly in the math notebook after solving.*

Answers

1. 48¢
2. 36¢
3. 60¢
4. 84¢
5. 54¢
6. 66¢
7. 34¢
8. 52¢
9. 70¢
10. 90¢

Multiplication with a Double-Digit Multiplier

(instructions to the parent)

Tip: Be sure to always call numbers by their proper name or place value; that is, 2 tens are referred to as "2 tens" and not simply "2", etc.

As you lead your child to multiplication with a double-digit multiplier, take your time and see if he is able to tell you what to do. This step is experimental. You may wish to use a piece of grid paper on which to write the problems so that each step remains visible rather than erasing on a slate.

1. Ask: If you collected 34 shells each day at the beach, how many shells would you have after 2 days?

2. On grid paper, write for him to solve:

$$\begin{array}{r} 34 \\ \times\ 2 \\ \hline \end{array}$$

3. Solution:

$$\begin{array}{r} 34 \\ \times\ 2 \\ \hline 68 \end{array}$$

4. Ask: If you collected 34 shells for 20 days, how many would you have?

5. Write:

$$\begin{array}{r} 34 \\ \times 20 \\ \hline \end{array}$$

6. Ask: What do you suppose we should do first? Be patient and see if your child figures out that there are 0 units with which to multiply the 4 units and 3 tens. Then he can move to the 2 tens and multiply 2 tens times 4 units and 2 tens times 3 tens. By adding the two products together, he solves the equation.

$$\begin{array}{rl} 34 & \\ \times 20 & \\ \hline 00 & = 34 \times 0 \textit{ units} \\ 680 & = 34 \times 2 \textit{ tens} \\ \hline 680 & \end{array}$$

You may want to see if your child knows the shortcut: that to multiply by 20 is simply multiplying by 2 and adding a zero. You might say something like, If you have 2 and want to change it to 20, what do we write after the 2? 0. So if we want to multiply by 20, what can we do? Multiply by 2 and add a zero.

$$\begin{array}{r} 34 \\ \times 20 \\ \hline 680 \end{array}$$

7. Ask: If you collected 34 shells for 22 days, how many would you have? Now see if he can tell you to first multiply by 2 units, then multiply by 20, and add the products. Since he has worked with addition of larger numbers he will likely be able to tell you this. If not, simply tell him that this is now worked just as an addition problem with larger numbers.

$$\begin{array}{r} 34 \\ \times\,22 \\ \hline 68 = 34 \times 2 \textit{ units} \\ 680 = 34 \times 2 \textit{ tens} \end{array}$$

$$\begin{array}{r} 34 \\ \times\;\; 22 \\ \hline {}_{1}68 \\ +680 \\ \hline 748 \end{array}$$

Tip: If your child begins with the tens rather than the units, that is fine as long as the proper place value is always kept. Many people find it easier to work the larger number first and the final product does not change, which is a great discovery. He may also realize on his own that no matter the order of the factors, or numbers being multiplied, the final product is the same.

Sums

1. There are 24 hours in a day. How many hours in 21 days?

2. How many hours in 23 days?

3. 25 cents for a gum ball. How much for 25 gum balls? (In problems of this sort, feel free to show where the dot, or decimal point, is put in the answer to signify dollars and cents. Decimals will be formally introduced in a later book.)

4. Lemonade costs 35¢ a glass at the lemonade stand. How much in all for 24 glasses of lemonade?

5. Pinwheels are 20¢ each. How much must you pay for 26 pinwheels? (Your student may realize on his own that no matter the order of the factors, or numbers being multiplied, the final product is the same. If he wants to work this as 26 x 20, let him.)

Answers

1. 504 hours
2. 552 hours
3. $6.25
4. $8.40
5. $5.20

Mixed Review

1. 7¢ + 3¢ + 4¢ + 2¢ = ?

2. Rachel collected 12 stickers. She gave 6 stickers to her younger sister. How many stickers does she have left?

3. $\begin{array}{r} 96 \\ +58 \\ \hline \end{array}$

4. 5 peaches are needed to bake a pie. How many peaches are needed to make 2 pies?

5. If you eat 2 apples a day for 7 days, how many apples have you eaten in all?

6. 2 x 4 =

7. $\begin{array}{r} 43 \\ -\ 8 \\ \hline \end{array}$

8. If gummy bears cost 2¢ and Gage bought 9, how much did he pay for all?

9. $\begin{array}{r} 386 \\ +193 \\ \hline \end{array}$

10. $\begin{array}{r} 587 \\ -236 \\ \hline \end{array}$

Answers

1. 16¢
2. 6 stickers
3. 154
4. 10 peaches
5. 14 apples
6. 8
7. 35
8. 18¢
9. 579
10. 351

Multiplication Table for 3

Tip: As a help, pull a Number Sentence Card out before the lesson begins so your student's work can continue uninterrupted if you get called away for a moment.

Each multiplication table for 3–6 will be learned by heart using the same steps, with the work being steady and deliberate. Be sure to work with the table in a variety of ways—not merely having your child parrot it in consecutive order.

Here is a simple outline of the steps you will follow. You may refer back to the detailed instructions on pages 197–210 as needed. As your student progresses you may determine if any steps may be omitted.

Step 1: Construct the table.
Step 2: Say the table and write it.

Step 3: Practice with Table Work (and review previous tables) until learned.
Step 4: Work with larger multiplication for that table.

Step 1: Construct the table.

Begin constructing a multiplication table for the number 3 while having the child determine how many times the number 3 is taken along with its corresponding product. You can create a table using objects first, if needed, then construct the written table or go immediately to the written table. Do whichever your child is ready for. (See pages 197–201 for detailed instructions of this step.)

1	2	3	4	5	6	7	8	9	10
3	3	3	3	3	3	3	3	3	3
3	6	9	12	15	18	21	24	27	30

Step 2: Say the table and write it.

Once the entire written multiplication table for 3 is constructed, give your child a few minutes to look at it and visualize it before having him say it through several times aloud; i.e., One 3 is 3, two 3s are 6, three 3s are 9, four 3s are 12, etc.

Erase a few numbers and ask your child to fill them in, then have your child say the table through again. Repeat that process several times.

Tip: When there is more than one child, the tables are learned individually and not in a chorus.

Next have your child construct a written table himself in his math notebook.

Step 3: Practice with Table Work until learned.

Ask a variety of questions involving the numbers in the table, allowing him to use the table for reference as needed. Continue giving a variety of questions daily until the table is learned. You may reuse the Table Work questions below or create your own similar variations.

1. Three 3s are?
2. Seven 3s are?
3. How much is 3 taken 4 times?
4. 3 taken 5 times is?
5. How many 3s in 15?
6. How many 3s in 18?
7. 3 taken 7 times is?
8. Lemonade at 3¢ for a cup means 4 cups will cost how much?
9. 3 taken 9 times is?
10. How many 3s in 24?
11. 3 taken 6 times is?
12. Chocolate candy costs 5¢ apiece, how much for 3 candies?
13. How many 3s in 27?
14. Stickers cost 3¢ each and you purchase 6. How much should you pay?
15. 3 x 3 is?
16. How many 3s in 30?
17. 3 taken 2 times is?

18. 3 x 4 is?

19. 3 taken 3 times is?

20. How many 3s in 12?

21. How many points did the football team score with 3 touchdowns at 6 points each?

22. 3 taken 10 times is?

23. How many 3s in 6?

24. 3 x 1 equals how many?

25. 3 taken 8 times is?

Answers

1. 9

Tip: Have the child give a fully worded answer to aid in retention of the facts; e.g., three 3s are 9 or 3 x 3 = 9.

2. 21
3. 12
4. 15
5. 5
6. 6
7. 21
8. 12¢
9. 27
10. 8
11. 18
12. 15¢
13. 9
14. 18¢
15. 9

16. 10
17. 6
18. 12
19. 9
20. 4
21. 18 points
22. 30
23. 2
24. 3
25. 24

Step 4: Work with larger multiplication by 3.

Objects used: dimes and pennies as needed.

Tip: See pages 204–209 for detailed instructions of this step.

1. Clementines are 23¢. How much would 3 clementines cost altogether?

2. Lemons cost 25¢ each. If you buy 3 lemons, how much must you pay?

3. A banana is 30¢. How much for 3 bananas?

4. Apples are 15¢. How much money is needed for 3 apples?

5. A lime costs 14¢. How much for 3? *To be written in math notebook after solving.*

6. We need 3 pears. At 21¢ each, how much must we give?

7. Oranges are priced at 33¢. How much for 3?

8. A grapefruit is 26¢. How much to buy 3 grapefruit?

9. Plums cost 12¢ each. How much is needed to purchase 3 plums?

10. If a mango costs 22¢, how much will 3 mangoes cost?

11. At 22¢ each, how much will 30 mangoes cost?

Tip: To be taken experimentally (#11–15).

12. Sample ice creams are 25¢ each. How much to try 34 different flavors?

13. Grapefruit are 26¢ each. How much for 33?

14. 31 × 31 =

15. 17 × 32 =

Answers:

1. 69¢
2. 75¢
3. 90¢
4. 45¢
5. 42¢; Child writes equation after solving.
6. 63¢
7. 99¢
8. 78¢
9. 36¢
10. 66¢
11. $6.60
12. $8.50
13. $8.58
14. 961
15. 544

Review with 3 and Previous Tables

End each lesson with five minutes of mental work, reviewing the previous tables learned. If needed, you can find additional mental work with 3 and previous tables on page 296.

1. 2 squirrels gather 6 acorns each. How many acorns did they gather all together?

2. At the farm, 2 cows give 4 gallons of milk each. How many total gallons did they produce?

3. A hen lays 3 eggs for 8 days. How many total eggs did she lay?

4. How many 2s in 4?

5. How many 3s in 9?

6. What will 4 apples at 2¢ each cost?

7. 3¢ x 7 =

8. 2¢ x 1 =

9. 2¢ x 5 =

10. How many 2s in 14?

11. The students sold 10 monster cookies for 2 dollars each. How much did they earn?

12. 6 children each pick 3 flowers. How many flowers all together?

13. 2 x 9 =

14. 3 x 3 =

15. 3 squirrels have 5 acorns each. How many acorns total?

16. 9 pencil cases each hold 3 pencils. How many pencils in all?

17. 3 kennels hold 2 dogs each. How many dogs in all?

18. 3¢ x 4 =

19. 2¢ x 8 =

20. How many 3s in 18?

Answers

1. 12 acorns
2. 8 gallons
3. 24 eggs
4. 2
5. 3
6. 8¢
7. 21¢
8. 2¢
9. 10¢
10. 7
11. $20
12. 18 flowers
13. 18
14. 9
15. 15 acorns
16. 27 pencils
17. 6 dogs
18. 12¢
19. 16¢
20. 6

Mixed Review

1. There are 5 cookies on a plate and 10 in the cookie jar. How many cookies in all?

2. Mr. Jamison bought 3 pounds of strawberries and 4 pounds of blueberries. How many pounds of berries had he in all?

3. Aviyah bought 4 embroidery needles and gave 2 away. How many needles does Aviyah have left?

4. If 3 soccer players score 2 goals each, how many goals are scored in all?

5.
$$\begin{array}{r} 5 \\ 3 \\ 2 \\ \underline{+6} \end{array}$$

6.
$$\begin{array}{r} 8 \\ 4 \\ 4 \\ \underline{+5} \end{array}$$

7. $\begin{array}{r} 54 \\ \underline{+23} \end{array}$ 8. $\begin{array}{r} 759 \\ \underline{+161} \end{array}$ 9. $\begin{array}{r} 33 \\ \underline{-18} \end{array}$ 10. $\begin{array}{r} 444 \\ \underline{-212} \end{array}$

Answers

1. 15 cookies
2. 7 pounds
3. 2 needles
4. 6 goals
5. 16
6. 21
7. 77
8. 920
9. 15
10. 232

Multiplication Table for 4

Tip: As a help, pull a Number Sentence Card out before the lesson begins so your student's work can continue uninterrupted if you get called away for a moment.

Step 1: Construct the table.

Begin constructing a multiplication table for the number 4 while having the child determine how many times the number 4 is taken along with its corresponding product.

1	2	3	4	5	6	7	8	9	10
4	4	4	4	4	4	4	4	4	4
4	8	12	16	20	24	28	32	36	40

Step 2: Say the table and write it.

Once the entire multiplication table for 4 is constructed, give your child a few minutes to look at it and visualize it before having him say it through several times aloud; i.e., One 4 is 4, two 4s are 8, three 4s are 12, etc.

Erase a few numbers and ask your child to fill them in, then have your child say the table through again. Repeat that process several times.

Have your child construct a written table himself in his math notebook.

Step 3: Practice with Table Work until learned.

Ask a variety of questions involving the numbers in the table, allowing your student to use the table for reference as needed. Continue giving a variety of questions daily until the table is learned. You may reuse the Table Work questions below or create your own similar variations.

1. How many 4s in 16?

2. 4 taken 5 times is?

3. There are 4 beats in a measure of music. How many beats in 4 measures?

4. How many 4s in 8?

5. How many 4s in 12?

6. Eight 4s are?

7. 4 taken 9 times is?

8. How many 4s in 28?

9. 4 taken 6 times is?

10. 4 taken 2 times is?

11. 4 x 9 is?

12. Find the cost of 7 rocks at 4¢ each.

13. How many 4s in 40?

14. Three 4s are?

15. 4 x 8 is?

16. How many 4s in 24?

17. 4 taken 1 time is?

18. How many 4s in 20?

19. 4 x 1 is?

20. 4 multiplied by 10 gives?

21. 4 taken 3 times is?

22. Five 4s are?

23. How many 4s in 32?

24. How many 4s in 36?

25. What is 4 multiplied by 9?

Answers

1. 4
2. 20
3. 16 beats
4. 2
5. 3
6. 32
7. 36
8. 7
9. 24
10. 8
11. 36
12. 28¢
13. 10
14. 12
15. 32
16. 6
17. 4
18. 5
19. 4
20. 40
21. 12
22. 20
23. 8

24. 9
25. 36

Step 4: Work with larger multiplication by 4.

Objects used: dimes and pennies as needed. Quarters to be used for #7.

Tip: Use the steps as laid out in Work with larger multiplication by 2 (pages 204–210) as necessary.

1. Paperclips are 13¢ each. How much for 4?
2. How much must you give for 4 bookmarks at 15¢ each?
3. A roll of washi tape is 22¢. How much for 4 rolls?
4. A sheet of paper costs 14¢. How much must you pay for 4 sheets?
5. Erasers are 23¢. How much for 4 erasers? *To be written in the math notebook after solving.*
6. If pencils are 17¢ each, how much would 4 pencils cost?
7. How much must you give for 4 markers at 25¢ apiece?
8. A ruler costs 20¢. How much would 4 rulers cost altogether?
9. Sticky notes are 16¢ for a pad. How much for 4 pads?
10. Paintbrushes are 18¢ each. How much for 4?
11. Paintbrushes are 18¢ each. How much for 40?

Tip: To be taken experimentally (#11–15).

12. How much for 42 sheets of paper at 14¢ each?

13. Puppy-shaped paperclips are 13¢ each. How much for 43 paperclips?

14. Owl-shaped paperclips are 22¢ each. How much for 42 of them?

15. $19 \times 44 =$

Answers

1. 52¢
2. 60¢
3. 88¢
4. 56¢
5. 92¢; Child writes equation after solving.
6. 68¢
7. $1.00
8. 80¢
9. 64¢
10. 72¢
11. $7.20
12. $5.88
13. $5.59
14. $9.24
15. 836

Review with 4 and Previous Tables

End each lesson with five minutes of mental work reviewing the previous tables learned. If needed, you can find additional mental work with 4 and previous tables on page 298.

1. A woodsman cut down 4 trees a day for 5 days. How many trees did he cut down in all?

2. 4 bags each contain 10 apples. How many apples all together?

3. 2¢ x 5 =

4. 4¢ x 8 =

5. 3¢ x 6 =

6. Jack saw 4 flocks of migrating birds with 7 birds in each flock. How many total birds did Jack see?

7. How many 2s in 16?

8. How many 3s in 12?

9. How much will 6 yards of cloth cost at 4 dollars a yard?

10. If one ice cream cone costs $3, what will 3 ice cream cones cost?

11. 2¢ x 7 =

12. 3¢ x 9 =

13. 4 ballplayers each have 2 baseball bats. How many bats in all?

14. 4 x 4 =

15. 3 x 2 =

16. How many 4s in 12?

17. Each shelf holds 4 books. How many books on 9 shelves?

18. 6 children each hold 2 balloons. How many balloons all together?

19. 3 x 7 =

20. 2 x 9 =

Answers

1. 20 trees

2. 40 apples
3. 10¢
4. 32¢
5. 18¢
6. 28 birds
7. 8
8. 4
9. $24
10. $9
11. 14¢
12. 27¢
13. 8 bats
14. 16
15. 6
16. 3
17. 36 books
18. 12 balloons
19. 21
20. 18

Mixed Review

1. $$\begin{array}{r} \$4.87 \\ \underline{+2.53} \end{array}$$

2. $$\begin{array}{r} \$5.96 \\ \underline{-2.84} \end{array}$$

3. How many 2s in 10?

4. $$\begin{array}{r} 648 \\ \underline{+294} \end{array}$$

5. $$\begin{array}{r} 57 \\ \underline{-16} \end{array}$$

6. 910
 -454

7. $2.93
 +1.48

8. $6.42
 -5.57

9. 2 robins each laid 3 eggs in their nests. How many eggs in all?

10. 284
 +243

Answers

1. $7.40
2. $3.12
3. 5
4. 942
5. 41
6. 456
7. $4.41
8. $ .85
9. 6 eggs
10. 527

Multiplication Table for 5

Step 1: Construct the table.

Begin constructing a multiplication table for the number 5 while having the child determine how many times the number 5 is taken along with its corresponding product.

1	2	3	4	5	6	7	8	9	10
5	5	5	5	5	5	5	5	5	5
5	10	15	20	25	30	35	40	45	50

Step 2: Say it through and write it.

Once the entire multiplication table for 5 is constructed, give your child a few minutes to look at it and visualize it before having him say it through several times aloud; i.e., One 5 is 5, two 5s are 10, three 5s are 15, four 5s are 20, etc.

Erase a few numbers and ask your child to fill them in, then have your child say the table through again. Repeat that process several times.

Have your child construct a written table himself in his math notebook.

Step 3: Practice with Table Work until learned.

Ask a variety of questions involving the numbers in the table, allowing him to use the table for reference as needed. Continue giving a variety of questions daily until the table is learned.

Tip: End each lesson with five minutes of mental work reviewing the previous tables learned.

1. 5 x 4 is?
2. 5 taken 8 times is?
3. How many 5s in 30?
4. There are 5 pigs in 2 fields. How many total pigs are there?
5. How many 5s in 40?
6. What is 5 multiplied by 8?
7. 5 taken 7 times is?

8. Nine 5s are?

9. How many 5s in 25?

10. 3 friends have 5 nickels each. How many total nickels do they have?

11. 5 x 1 is?

12. How many 5s in 50?

13. Three 5s are?

14. 5 x 5 is?

15. How many 5s in 35?

16. What is 5 taken 6 times?

17. How many 5s in 20?

18. Gum balls are 5¢ apiece. How much for 5 gum balls?

19. 4 children picked 5 apples each. How many apples did they pick all together?

20. 5 taken 2 times is?

21. How many 5s in 15?

22. How much do 9 pencils cost at 5¢ each?

23. 5 x 6 is?

24. How many 5s in 10?

25. What is 5 taken 7 times?

Answers

1. 20
2. 40
3. 6
4. 10 pigs
5. 8
6. 40
7. 35
8. 45
9. 5
10. 15 nickels
11. 5
12. 10
13. 15
14. 25
15. 7
16. 30
17. 4
18. 25¢
19. 20 apples
20. 10
21. 3
22. 45¢
23. 30
24. 2
25. 35

Step 4: Work with larger multiplication by 5.

Objects used: dollars, dimes, and pennies as needed. Quarters are used for #4.

1. Plastic beads cost 15¢ each. How many for 5?

2. Wooden beads are 20¢ each. How much must you give for 5?

3. Felt is 19¢ a square. How much for 5 squares?

4. Glitter is 25¢ a tube. How much for 5 tubes?

5. Pipe cleaners are 28¢ each. How much for 5?

6. Googly eyes are 26¢ a pair. How much for 5 pair?

7. Pompoms are 17¢ each. How much must you give for 5 of them?

8. Feathers cost 24¢. How much for 5?

9. Glue sticks cost $1.03 each. How much for 5 sticks? *To be written by you on the slate, seeing if your child can tell where each number in the product is written.*

10. Thread is $1.10 a spool. How much should you give for 5 spools? *To be written by your child on his slate and then in his math notebook after solving.*

11. At 15¢ each, how much for 50 beads?

Tip: To be taken experimentally (#11–15).

12. How much for 55 beads at 15¢ apiece?

13. At 19¢ apiece for felt squares, how much for 52 pieces of felt?

14. Pompoms are 17¢ each. How much for 53?

15. 20 × 50 =

Answers

1. 75¢
2. $1.00
3. 95¢
4. $1.25

5. $1.40
6. $1.30
7. 85¢
8. $1.20
9. $5.15
10. $5.50
11. $7.50
12. $8.25
13. $9.88
14. $9.01
15. 1000

Review with 5 and Previous Tables

End each lesson with five minutes of mental work reviewing the previous tables learned. If needed, you can find additional mental work with 5 and previous tables on page 300.

1. At $10 apiece, what will 4 tickets cost?

2. Lyla has 9 crayons and Gabe has 5 times as many. How many crayons does Gabe have?

3. 2 x 8 is?

4. If cotton candy costs $3, how much for 2 cotton candies?

5. What is the cost of 7 yards of ribbon at $4 a yard?

6. How many 3s in 21?

7. How many 4s in 16?

8. A boy walks 3 miles an hour. How far will he walk in 3 hours?

9. 5¢ x 6 =

10. 4¢ x 9 =

11. 3¢ x 8 =

12. An apple orchard has 8 rows of trees with 5 trees in each row. How many apple trees in all?

13. 4 spiders with 8 legs each. How many legs in all?

14. 2 x 7 =

15. At $3 each, how many cupcakes can you buy for $12?

16. 5¢ x 10 =

17. A gum ball is 1¢. How much for 5 gum balls?

18. 5 x 5 =

19. 3 x 9 =

20. 4 x 6 =

Answers

1. $40
2. 45 crayons
3. 16
4. $6
5. $28
6. 7
7. 4
8. 9 miles
9. 30¢
10. 36¢
11. 24¢
12. 40 trees

13. 32 legs
14. 14
15. 4 cupcakes
16. 50¢
17. 5¢
18. 25
19. 27
20. 24

Mixed Review

1. Iris biked 4 miles to the post office and 4 miles back to her home. How many miles did Iris bike in total?

2. Max read 6 chapters of his new book yesterday and 9 chapters today. How many chapters did he read in all?

3. We have 5 pounds of sugar in the house but will need 8 pounds of sugar for our baking day. How many more pounds of sugar do we need?

4. 19 rabbits were in a field when 9 hopped away. How many rabbits remained?

5. $$\begin{array}{r} 26 \\ +\ 9 \\ \hline \end{array}$$

6. $$\begin{array}{r} 54 \\ +28 \\ \hline \end{array}$$

7. $$\begin{array}{r} 432 \\ +386 \\ \hline \end{array}$$

8. $$\begin{array}{r} 37 \\ -12 \\ \hline \end{array}$$

9. 651
 - 9

10. 832
 -152

Answers

1. 8 miles
2. 15 chapters
3. 3 pounds
4. 10 rabbits
5. 35
6. 82
7. 818
8. 25
9. 642
10. 680

Multiplication Table for 6

Step 1: Construct the table.

Begin constructing a multiplication table for the number 6 while having the child determine how many times the number 6 is taken along with its corresponding product.

1	2	3	4	5	6	7	8	9	10
6	6	6	6	6	6	6	6	6	6
6	12	18	24	30	36	42	48	54	60

Step 2: Say it through and write it.

Once the entire multiplication table for 6 is constructed, give your child a few minutes to look at it and visualize it before having him say it through several times aloud; i.e., One 6 is 6, two 6s are 12, three 6s are 18, four 6s are 24, etc.

Erase a few numbers and ask your child to fill them in, then have your child say the table through again. Repeat that process several times.

Have your child construct a written table himself in his math notebook.

Step 3: Practice with Table Work until learned.

Ask a variety of questions involving the numbers in the table, allowing him to use the table for reference as needed. Continue giving a variety of questions daily until the table is learned.

1. 6 taken 5 times is?
2. How many 6s in 48?
3. 6 x 3 is?
4. 2 shoes make a pair. How many shoes in 6 pair?
5. How many 6s in 54?
6. Seven 6s are?
7. How many 6s in 42?
8. 6 x 4 is?
9. 6 children each have $1. How many dollars in all?
10. How many 6s in 36?
11. 6 x 9 is?
12. How many 6s in 60?

13. Paper costs 6¢ a sheet. If Thomas buys 4 sheets, how much would he spend?

14. 6 x 5 is?

15. 6 taken 6 times is?

16. How many 6s in 18?

17. Six 6s are?

18. 6 x 8 is?

19. How many 6s in 30?

20. 6 x 10 is?

21. How many 6s in 24?

22. 6 taken twice is?

23. 6 x 7 equals?

24. Three 6s are?

25. How many 6s in 12?

Answers

1. 30
2. 8
3. 18
4. 12 shoes
5. 9
6. 42
7. 7
8. 24

9. $6
10. 6
11. 54
12. 10
13. 24¢
14. 30
15. 36
16. 3
17. 36
18. 48
19. 5
20. 60
21. 4
22. 12
23. 42
24. 18
25. 2

Step 4: Work with larger multiplication by 6.

Objects used: craft sticks and bundles as needed

1. There are 15 apples trees in a row. How many trees in 6 rows?

2. There are 12 eggs in a dozen. How many eggs in 6 dozen?

3. Jenny placed 18 tulips in a bouquet. How many tulips in 6 bouquets?

4. There are 21 birds on a wire. How many birds on 6 wires?

5. James collected 24 pinecones a day for 6 days. How many pinecones did he collect in all? *To be written in the math notebook after solving.*

6. There are 26 beads on each bracelet. How many beads on 6 bracelets?

7. There are 22 roses on each bush. How many roses on 6 bushes?

8. Ashley collected 27 shells each day on the beach. How many had she collected in 6 days?

9. The pet store has 19 fish in each tank. How many fish are in 6 tanks?

10. A farmer keeps 25 sheep in each field. How many sheep in 6 fields? *To be written in the math notebook after solving.*

11. At the orchard there are 15 apple trees in a row. How many in 60 rows?

Tip: To be taken experimentally (#11–15).

12. 12 eggs in a dozen. How many eggs in 65 dozen?

13. 15 roses in a bouquet. How many roses in 61 bouquets?

14. 14 × 66 =

15. 13 x 62 =

Answers

1. 90 trees
2. 72 eggs
3. 108 tulips
4. 126 birds
5. 144 pinecones
6. 156 beads
7. 132 roses
8. 162 shells
9. 114 fish
10. 150 sheep
11. 900 trees
12. 780 eggs
13. 915 roses
14. 924
15. 806

Review with 6 and Previous Tables

End each lesson with five minutes of mental work reviewing the previous tables learned. If needed, you can find additional mental work with 6 and previous tables on page 302.

1. If 3 people each have a 5-gallon bucket of water, how many gallons of water in all?

2. How much are 9 rubber ducks at $3 each?

3. How much are 4 gum balls at 5¢ each?

4. 6 students each read 4 books. How many books are read all together?

5. How much are 8 stickers at 4¢ each?

6. 2 x 9 =

7. 4 x 4 =

8. 5 students each read 8 lines during the Shakespeare lesson. How many lines were read in all?

9. Each bluet flower has 4 petals. How many petals have 9 flowers all together?

10. A farmer has 3 pens with 8 goats in each. How many goats in all?

11. How much are 3 hairbrushes at $4 each?

12. 6 children on the bridge. Each drops 2 sticks into the water. How many sticks are dropped into the water?

13. How many 3s in 21?

14. How many 4s in 40?

15. Each child is allowed 3 cookies. How many cookies for 2 children?

16. How many cookies for 3 children?

17. 5 x 9 =

18. 6 x 7 =

19. How many 5s in 35?

20. How many 4s in 28?

Answers

1. 15 gallons
2. $27
3. 20¢
4. 24 books
5. 32¢
6. 18
7. 16
8. 40 lines
9. 36 petals
10. 24 goats
11. $12
12. 12 sticks
13. 7
14. 10
15. 6 cookies
16. 9 cookies
17. 45
18. 42
19. 7
20. 7

Mixed Review

1. Stewart rowed 5 miles across the lake and 5 miles back. How many miles did Stewart row his boat?

2. Kayla braided 7 bookmarks yesterday and 8 bookmarks today. How many bookmarks did she braid in all?

3. The Andrews have 4 bags of cement for their project but will need 8 bags. How many more bags of cement do they need?

4. 17 turtles were on a log when 9 went back into the pond. How many turtles remained?

5. $\begin{array}{r} 33 \\ +\ 5 \\ \hline \end{array}$

6. $\begin{array}{r} 62 \\ +18 \\ \hline \end{array}$

7. $\begin{array}{r} 324 \\ +368 \\ \hline \end{array}$

8. $\begin{array}{r} 45 \\ -13 \\ \hline \end{array}$

9. $\begin{array}{r} 763 \\ -\ \ 9 \\ \hline \end{array}$

10. $\begin{array}{r} 942 \\ -277 \\ \hline \end{array}$

Answers

1. 10 miles
2. 15 bookmarks
3. 4 bags
4. 8 turtles

5. 38
6. 80
7. 692
8. 32
9. 754
10. 665

Tip: In this book, Book 2, the child learns multiplication tables up through 6. The tables for 7 through 10 will be presented in Book 3.

Multiplication by 10s, 100s, 1000s

Objects used: dimes and dollars

1. Ask: How many pennies in a dime? (10)

2. Write *10* on the slate in the upper right corner.

3. Take out a number of dimes and have your child count by 10s to 100.

4. Ask: How many 10s make 100? (10)

5. Write *100* underneath the *10*, being sure to keep proper place value.

6. Say: Using our $1 bill to represent 100 pennies, count by 100s to 1000.

7. Ask: How many 100s in 1000? (10)

8. Write *1000* beneath the *100*, keeping proper place value.

9. Say: Now we will multiply numbers by 10.

10. Write a *5* on the slate (about midpoint on the left) and ask: What is 5 taken 10 times? (50)

11. Write a *0* after the *5* to represent *50.*

12. Write a *2* on the slate and ask: And if I had 2 taken 10 times, what should I write after the 2? (0)

13. If this is clear to your child, you may continue to step #14, but if not, repeat steps 10 and 11 a few more times using the numbers *3* and *6.*

14. Erasing everything from steps #10 and #11, write down a *5* and then write *50* underneath it, keeping proper place value.

15. Ask: What would be 5 times 100? (500)

16. Write the *5* in the hundreds place below the *50* and ask: What should come after the 5? (Two 0s)

17. Ask: And what if we multiplied 5 by 1000? (5000)

18. Write the *5* in the thousands place underneath *500* and ask: How many zeroes should come after the 5? (Three 0s)

19. If this is clear, you may continue to the next step. If necessary, repeat steps 14–18 with another number, emphasizing how many 0s are added to the number in each product or answer.

20. Now you will give a variety of numbers to have your child multiply by 10, 100, and 1000. Do not give them always in consecutive order; rather, emphasize the number of 0s added to the number to make each product or answer. Thus, you will write each number on the slate and ask how many zeroes are needed.

21. Ask: What is 2 times 1000? (2000)

22. Ask: How many zeroes after the 2? (Three 0s)

23. Ask: What is 2 multiplied by 100? (200)

24. Ask: How many zeroes after the 2? (Two 0s)

25. Ask: What is 9 taken 10 times? (90)

26. Ask: How many zeroes after the 9? (One 0)

27. Ask: What is 9 times 1000? (9000)

28. Ask: How many zeroes after the 9? (Three 0s)

29. Ask: What is 7 multiplied by 100? (700)

30. Ask: How many zeroes after the 7? (Two 0s)

31. Ask: What is 7 times 10? (70)

32. Ask: How many zeroes after the 7? (One 0)

33. Ask: If a number is multiplied by 1000, how many zeroes are added to it? (Three)

34. Ask: If a number is multiplied by 10, how many zeroes are added to it? (One)

35. Ask: And if a number is multiplied by 100, how many zeroes are added to it? (Two)

Pure Number

To be given orally.

1. 3 x 100 =

2. 8 x 1000 =

3. 2 x 10 =

4. 6 x 1000 =

5. 10 x 10 =

6. 10 x 100 =

7. 4 x 10 =

8. 5 x 100 =

9. 7 x 1000 =

10. 2 x 100 =

Answers

1. 300
2. 8000
3. 20
4. 6000
5. 100
6. 1000
7. 40
8. 500
9. 7000
10. 200

Review

1. Along the beach, Christy finds 16 shells in the morning, 12 in the afternoon, and 17 in the evening. How many shells did Christy find in all?

2. Leo practiced his violin for 15 minutes yesterday and then practiced 35 minutes today. How much total time did Leo practice?

3. Rachel picked 4 flowers for her mom. If she picked 19 flowers earlier, how many flowers did she pick in all?

4. Lyla is 52 inches tall and Evienne is 47 inches tall. What is the difference between their heights (or how much taller is Lyla than Evienne)?

5. The Logans have seen 53 animals at the zoo, and there are 26 animals left to see. How many total animals are at the zoo?

6. Lainie bought 6 pencils, 4 notebooks, and 25 paper clips. How many supplies did he purchase?

7. David practiced soccer for 43 minutes yesterday and for 54 minutes today. How many minutes did he spend practicing all together?

8. There are 96 books on the bookshelf. Robert adds 5 more books. How many total books are on the shelf?

9. There are 4 chipmunks, 7 rabbits, and 3 deer in the yard. How many total animals?

10. Hudson buys a game for $14. He gives the clerk $20. How much change should he get?

Answers

1. 45 shells
2. 50 minutes
3. 23 flowers
4. 5 inches
5. 79 animals
6. 35 supplies
7. 97 minutes
8. 101 books
9. 14 animals
10. $6

Formal Introduction of Division

We are at the point in our journey where division is formally introduced. Your child will have already experienced a natural overlap with division, having seen that "half of 10 is 5" and "four 3s are contained in 12" through his work with numbers, objects, and the multiplication tables.

In this section we will be communicating a different idea of division—the idea of continuous subtraction, or *sharing*—as well as presenting the symbol for division and its meaning. (Note: Terminology and a second idea of division as fractional parts will be presented in Book 3).

You may wish to view "Division" in *Charlotte Mason Living Math: A Guided Journey* for demonstrations. That session will extend beyond what you will cover here, in Book 2, and give you a sneak peek at what is coming in Book 3.

Division as Continuous Subtraction

Sample *(instructions to the parent)*

Objects used: candies, nuts, or something else that is easily shared

To begin, we will give the idea of continuous subtraction through simple sharing exercises. You will be writing on the slate until you instruct your child to do so.

1. Say: We have 12 candies to share so that each friend receives 4. Let's begin by taking away the first 4 candies for the first child and putting them in a pile.

   Write on the slate: 12 – 4

2. Say: We still have 8 candies left to share, so take out the next 4 candies to put into another pile, or group, for another friend.

   Notate the subtraction of another 4 on the slate: 12 – 4 – 4

3. Say: Now, take out the next 4 candies.

   Write that step on the slate. It should look like this: 12 – 4 – 4 – 4

   Say: We have no more candies to share. How many piles do we have? (3 piles)

   Ask: So how many friends would receive 4 candies? (3 friends)

4. Now give another problem. Ask: If you have 16 candies and share them into piles of 4, how many children would have a pile?

   Your child will subtract or divide into piles the 16 candies by 4s until there are no more candies left for sharing. Answer: 4 children.

   Write on the board: 16 – 4 – 4 – 4 – 4

   Explain: This sharing into piles, or groups, of 4 is called "division by 4."

5. Give a few more problems of this nature. Ask: If you have 9 candies, how many children could get a share of 3 each?

He divides his candies into piles of 3 and answers: 3 children.

Again explain: This sharing into piles of 3 is called "division by 3."

Now say something like: If I ask you to divide 8 by 2s, what would you do?

He will probably be able to answer something like: Divide it into piles, or groups, of 2.

Ask: So 8 divided by 2 is?

He may work the problem using the candies or may answer without their use: We would have 4 piles because there are 4 twos in 8.

6. Say while writing: If we write 8 – 2 – 2 – 2 – 2 we see we have four 2s, but we can write it in a shorter way to show "division by 2."

The sign ÷ means "divided by." So if we say 8 divided by 2 equals 4, we write it like this: 8 ÷ 2 = 4, showing the number on the left is divided by the number on the right.

7. Ask: If you had 6 candies and divided them into groups of 3s, how many groups would you have? (2 groups)

8. Ask: How would it be written using the division symbol? Child should write: 6 ÷ 3 = 2

9. Ask: And how would it be read? (Six divided by three equals two.)

Sums

Now give more questions of the same type. Each time have your student use manipulatives as necessary to solve the problem, then write the problem on the slate, then read the entire equation. He should solve, write, and read for each question.

1. If you divide 10 candies into piles of 5, how many piles would you have?

2. How many if you divide 12 candies by 2?

3. How many if you divide 18 candies by 6?

4. If you divide 9 candies so that each person receives 3, how many people will get candy?

5. If 15 candies are divided by 3, how many friends receive a share?

6. How many if you divide 8 candies by 4?

7. 20 candies divided into groups of 5 equals how many groups?

Answers:

1. 2 piles, because $10 \div 5 = 2$
2. 6 piles, because $12 \div 2 = 6$
3. 3 piles, because $18 \div 6 = 3$
4. 3 people, because $9 \div 3 = 3$
5. 5 friends, because $15 \div 3 = 5$
6. 2 friends, because $8 \div 4 = 2$
7. 4 groups, because $20 \div 5 = 4$

Division with Money

(instructions for the parent)

Objects used: coins

Now you will work a good number of money sums with your child, having him work from left to right with dimes worked first–exchanging dimes for pennies if need be. Have your child speak each problem as it is being worked. Once a problem is solved, he may write it on the slate or in his math notebook.

Tip: As the habit of attention is cultivated, you may be surprised at how many numbers a child is able to keep in his head. If the habit is still being nurtured, don't allow this to be a point of frustration. You may write the number sentences on the slate if needed while you continue to encourage him to speak through what he is doing.

1. How many gum balls at 5¢ each can you buy with 20¢?

   Child talks as he solves the equation: *20 ÷ 5 = 4*

Tip: If the child is using 2 dimes, lead him through with questioning if need be, but by now he is probably so used to exchanging coins that he will easily see that he must exchange his dimes for pennies. If not, simply say something like: You cannot divide 2 dimes into groups of 5. What should you do? That's right, exchange 2 dimes for 20 pennies and then divide by 5.

2. How many gum balls at 5¢ each can you buy for 30¢?

3. How many candies at 2¢ each can you buy for 20¢?

4. How many candies at 5¢ each can you buy for 50¢?

5. How many stickers at 4¢ each can you buy for 16¢? *To be written after solved.*

6. How many stickers at 4¢ each can you buy for 12¢?

7. How many pencils at 5¢ each can you buy for 35¢?

8. How many pencils at 3¢ each can you buy for 24¢?

9. How many ribbons at 6¢ each can you buy for 18¢?

10. How many ribbons at 6¢ each can you buy for 24¢? *To be written after solved.*

11. 12¢ divided by 3?

12. 30¢ divided by 6?

13. 24¢ divided by 4?

14. 21¢ divided by 3?

15. 16¢ divided by 2? *To be written after solved.*

16. 14¢ divided by 2?

17. 45¢ divided by 5?

18. 27¢ divided by 3?

19. 48¢ divided by 6?

20. 40¢ divided by 4? *To be written after solved.*

Answers

1. 4 gum balls

Tip: Be sure to require fully worded answers; for example 20 ÷ 5 = 4

2. 6 gum balls
3. 10 candies
4. 10 candies
5. 4 stickers
6. 3 stickers
7. 7 pencils
8. 8 pencils
9. 3 ribbons
10. 4 ribbons
11. 4¢
12. 5¢
13. 6¢
14. 7¢
15. 8¢

16. 7¢
17. 9¢
18. 9¢
19. 8¢
20. 10¢

Longer Division with Objects

Objects used: dollars and coins

Next, longer division problems are worked in only the concrete, using dollar bills and coins, by having the child divide the money into equal groups.

1. What is $2.64 divided evenly among 2 people?

Tip: Be sure to always keep the position of the dollars, dimes, and pennies in their proper order and have your child first divide the dollars, then the dimes, and finally, the pennies.

2. What is $4.84 divided by 2 people?

3. What is $6.20 divided among 2?

4. What is $3.96 divided by 3?

5. What is $6.03 divided by 3?

6. What is $8.84 divided by 4 people?

7. What is $4.28 divided by 4 people?

8. What is $10.25 divided among 5?

9. What is $5.55 divided by 5?

10. What is \$6.12 divided between 6 people?

Answers

1. \$1.32

Tip: Be sure to require fully worded answers; for example, \$2.64 ÷ 2 = \$1.32

2. \$2.42
3. \$3.10
4. \$1.32
5. \$2.01
6. \$2.21
7. \$1.07
8. \$2.05
9. \$1.11
10. \$1.02

Review Addition and Subtraction

1. Sarina received 7 cards for her birthday and the postman brought 4 more. How many total cards did she receive for her birthday?

2. Daniel sees 5 frogs in the pond, 7 frogs on lily pads, and 11 more frogs along the edge. How many total frogs does Daniel see?

3. Ruth knit 5 rows on her scarf yesterday, 12 rows this morning, and 7 more rows in the evening. How many rows did she knit in total?

4. Macy must practice her violin for 25 minutes. She has practiced for 18 minutes. How many more minutes must Macy practice?

5. Hank has 8 balloons. His younger brother also has 8 balloons and his older brother has 4 balloons. How many balloons do the boys have in all?

6. Carrie picked 13 zucchini at the beginning of the week, 5 zucchini yesterday, and 6 today. What is the total number of zucchini Carrie harvested this week?

7. Barton is 18 years old and Ren is 9 years old. How many years younger is Ren than Barton (or what is the difference in their ages)?

8. Molly gathered 8 pink roses, 6 white roses, and 9 red roses. How many roses has Molly gathered in all?

9. Gabe planted 12 apple trees, 12 pear trees, and 18 peach trees. How many trees did he plant in all?

10. There are 26 donuts and 19 muffins on the table. How many fewer muffins are there than donuts?

Answers

1. 11 cards
2. 23 frogs
3. 24 rows
4. 7 minutes
5. 20 balloons
6. 24 zucchini
7. 9 years
8. 23 roses
9. 42 trees
10. 7 muffins

Short Division with New Notation

When the divisor is a simple, or a 1-digit number, what we term *Short Division* can be used, going through the sequence of steps mentally. Craft sticks in both ten bundles and units are used to demonstrate the problems so the child sees the *why* behind keeping the place value in this new style of notation. For that reason, the first problem uses a larger dividend –that is, the number being divided into– before working with numbers that fall within the multiplication tables already studied. Remember, there is no need to rush. Allow your child plenty of time to work thoughtfully and neatly.

Sample *(instructions for the parent)*

Objects used: craft sticks in ten bundles and units

1. Have your child make 24 using 2 ten bundles and 4 units.
   Write: 24 ÷ 2

   Explain: You already know this means the number of 2s in 24, or 24 divided by 2. We can also write 24 ÷ 2 like this:

   $2\overline{)24}$

   Have the child divide the 2 ten bundles and 4 units into two groups, then say–while pointing to each number and showing where each is written: 2 tens divided by 2 = 1 ten (write the *1* in the tens place above the *2* in *24*), and 4 units divided by 2 = 2 units (write the *2* in the units place above the *4*). The answer to 24 divided by 2 is 12.

2. Ask your student to make 12 using a ten bundle and units.
   Write 12 ÷ 2

   Explain: This means the number of 2s in 12 or 12 divided by 2. We can also write 12 ÷ 2 like this:

   $2\overline{)12}$

   Now ask the child to divide 12 into 2 groups, starting with the 1 ten bundle. He cannot divide it unless he breaks it apart and turns it into units. Explain that, for this reason, the answer, *6*, is written above the *2* in *12*.

Sums

Now give your child some short division problems, asking him to show you how to write the first few problems on his slate using this new way of writing division problems. Remember, this is something new so you may show how again if necessary. These are small enough problems that you may encourage your student to find the answers mentally, getting the sticks out if needed for solving or if proper place value is not being kept in notation.

1. 10 ÷ 2 = *He may write this problem on his slate.*

2. 9 ÷ 3 = *He may write this problem on his slate.*

3. 12 ÷ 2 =

4. 12 ÷ 3 =

5. 12 ÷ 6 = *He may write this problem on his slate.*

6. 18 ÷ 6 =

7. 14 ÷ 2 =

8. 16 ÷ 2 =

9. 15 ÷ 5 =

10. 20 ÷ 4 = *He may write this problem on his slate.*

Answers

1. 5
2. 3
3. 6
4. 4
5. 2
6. 3
7. 7
8. 8

9. 3
10. 5

Short Division with Money

Now you may give your child a mix of problems for additional practice and to help build speed and accuracy. You may have him write some problems on his slate or math notebook every so often as a problem is solved. The problems are still not to be written before being solved.

1. How many rubber bands at 3¢ each can you buy for 9¢?

2. How many rubber bands at 3¢ each can you buy for 21¢?

3. How many gum balls at 5¢ each can you buy for 45¢?

4. How many gum balls at 5¢ each can you buy for 40¢?

5. How many candies at 4¢ each can you buy for 40¢?

6. How many candies at 6¢ each can you buy for 54¢?

7. How many stickers at 4¢ each can you buy for 20¢?

8. How many stickers at 4¢ each can you buy for 16¢?

9. How many pencils at 4¢ each can you buy for 28¢?

10. How many pencils at 6¢ each can you buy for 42¢?

Answers

1. 3 rubber bands

Tip: Be sure to require fully worded answers; for example, 9 ÷ 3 = 3

2. 7 rubber bands
3. 9 gum balls
4. 8 gum balls

5. 10 candies
6. 9 candies
7. 5 stickers
8. 4 stickers
9. 7 pencils
10. 7 pencils

Short Division Mixed Review

1. Rachel puts on a puppet show with her three sisters. They have a total of 16 puppets to divide between them. How many puppets do each of the 4 girls get?

2. Leo makes a tool bench and uses 32 total nails on the 4 legs. How many nails are used on each bench leg?

3. Andrew and Mark picked 18 pounds of blueberries. If they share their harvest evenly, how much will each boy get?

4. Jack practiced his piano for 50 minutes over 5 days. How many minutes did he practice each day?

5. Shepherd purchased 5 stickers for a total of 35¢. How much did each sticker cost?

6. Gum balls are 4¢ apiece. If Maddy paid 24¢ total, how many gum balls did she purchase?

7. Mary earned 40¢ for watering 4 plants. How much did she earn for each plant?

8. Ruthie purchased cardstock paper for 48¢. If each sheet of cardstock cost 6¢, how many sheets was Ruthie able to purchase?

9. Daniel paid 50¢ for 5 kazoo lessons. How much did each lesson cost?

10. Annalise had 12 cookies to share among 4 children. How many cookies did each child get?

11. 16 ÷ 2 =

12. 45 ÷ 5 =

13. 9 ÷ 3 =

14. 20 ÷ 5 =

15. 8 ÷ 2 =

16. 28 ÷ 4 =

17. 14 ÷ 2 =

18. 24 ÷ 3 =

19. 15 ÷ 3 =

20. 10 ÷ 5 =

Answers

1. 4 puppets

Tip: Be sure to require fully worded answers.

2. 8 nails
3. 9 pounds
4. 10 minutes
5. 7¢
6. 6 gum balls
7. 10¢
8. 8 sheets
9. 10¢
10. 3 cookies

11. 8
12. 9
13. 3
14. 4
15. 4
16. 7
17. 7
18. 8
19. 5
20. 2

Review Multiplication and Division

1. Peter baked 24 muffins to share equally between 2 neighbors. How many muffins does each neighbor receive?

2. Lyla knit 3 baby hats each day for 5 days. How many hats did Lyla knit in all?

3. 4 boys each have 5 trading cards. How many cards do they have in all?

4. A farmer puts half his flock of 18 sheep in a pasture. How many sheep are in the pasture?

5. The Harley's grandfather gives $50 to be distributed equally to his 5 grandchildren. How many dollars does each child receive?

6. Earning $6 an hour for yard work, how many hours must Collette work to earn $54?

7. Collette has earned $42 for yard work. If she is paid $6 an hour, how many hours has she worked?

8. If Abby can fit 6 books on each shelf, how many shelves are needed to hold 24 books?

9. Caroline reads 2 picture books a day for 4 days. How many does she read in all?

10. There are 4 slides at the playground and Macy rides each slide 7 times. How many times does she slide in all?

Answers

1. 12 muffins
2. 15 hats
3. 20 cards
4. 9 sheep
5. $10
6. 9 hours
7. 7 hours
8. 4 shelves
9. 8 books
10. 28 times

Extra Mental Arithmetic

The lessons in this book provide many problems for use in mental arithmetic, but here are more to use as needed and to help get your creativity flowing. You will find extra practice for

Addition & Subtraction with 3 and Previous Tables
Addition & Subtraction with 4 and Previous Tables
Addition & Subtraction with 5 and Previous Tables
Addition & Subtraction with 6 and Previous Tables
Addition & Subtraction with 7 and Previous Tables
Addition & Subtraction with 8 and Previous Tables
Addition & Subtraction with 9 and Previous Tables
Addition & Subtraction with 10 and Previous Tables
Mixed Simple Addition & Subtraction
Mixed Addition & Subtraction Money Sums
Mixed Addition & Subtraction with 3 or More Figures
Mixed Addition & Subtraction in Pure Number
Combined Addition & Subtraction with 3 or More Figures
Combined Addition & Subtraction with 5 or More Figures
Multiplication & Division with 3 and Previous Tables
Multiplication & Division with 4 and Previous Tables
Multiplication & Division with 5 and Previous Tables
Multiplication & Division with 6 and Previous Tables
Mixed Multiplication & Division
Multiplication with Money Problems
Division with Money Problems
Multiplication in Pure Number
Division in Pure Number
Combined Multiplication & Division in Pure Number

Tip: Twist and turn the questions to keep them interesting.

Tip: Adjust the speed to strike a balance between going slowly enough that your child does not panic but rapidly enough to fix his attention.

Addition & Subtraction with 3 and Previous Tables

1. 3 cats and 4 kittens. How many all together?

2. 3 dogs and 6 puppies. How many in all?

3. 7 cupcakes on a plate. If 2 get eaten, how many remain?

4. 2 apple pies and 3 pumpkin pies equal how many pies?

5. There were 8 trees and 3 were chopped down. How many are left now?

6. There were 5 lamps in a room. If 3 were turned on, how many were off?

7. A boy has 2 pockets on his shirt and 4 on his jeans. How many pockets has he?

8. A girl had 10 bracelets but lost 2. How much had she left?

9. A boy had 8¢ and bought a gum ball for 2¢. How much had he remaining?

10. 7 birds at the feeder when 1 flew away. How many birds remained?

11. 1 + 3 =

12. 2 + 6 =

13. 3 + 3 =

14. 9 – 2 =

15. 10 – 1 =

16. 11 – 1 =

17. 2 + 7 =

18. 6 – 3 =

19. 4 – 2 =

20. 3 + 8 =

Answers

1. 7 cats
2. 9 dogs
3. 5 cupcakes
4. 5 pies
5. 5 trees
6. 2 lamps
7. 6 pockets
8. 8 jelly bands
9. 6¢
10. 6 birds
11. 4
12. 8
13. 6
14. 7
15. 9
16. 10
17. 9
18. 3
19. 2
20. 11

Addition & Subtraction with 4 and Previous Tables

1. 2 horses in a meadow. How many legs in all?

2. There were 8 geese on a pond when 4 flew away. How many geese were left?

3. A boy had 9 pennies and lost 3. How many had he remaining?

4. His father gave him 4 more pennies. How many had he now? (6 + 4)

5. A woman had 4 notebooks but wants 6. How many more does she need?

6. A boy saw 4 cardinals and 6 bluejays. How many birds in all?

7. There were 13 mushrooms by the tree. If 3 were picked, how many remained?

8. Mr. Jones put 4 hotdogs on the grill. He then put 7 more hotdogs on the grill. How many in all?

9. 12 muffins on a tray. If 2 are eaten, how many remain?

10. There were 10 children on the playground. 3 children left. How many remained?

11. 4 + 5 =

12. 3 + 8 =

13. 12 – 3 =

14. 6 – 2 =

15. 1 + 10 =

16. 4 + 4 =

17. 2 + 5 =

18. 5 – 4 =

19. 4 – 0 =

20. 2 + 9 =

Answers

1. 8 legs
2. 4 geese
3. 6 pennies
4. 10 pennies
5. 2 notebooks
6. 10 birds
7. 10 mushrooms
8. 11 hotdogs
9. 10 muffins
10. 7 children
11. 9
12. 11
13. 9
14. 4
15. 11
16. 8
17. 7
18. 1
19. 4
20. 11

Addition & Subtraction with 5 and Previous Tables

1. Sophie cut 2 roses off a bush and 5 roses off another. How many had she in all?

2. 10 robins on a hill when 5 flew away. How many robins remained?

3. Sammy had a toy truck with 8 wheels. 5 wheels fell off. How many remained on?

4. 5 balls of red yarn and 6 balls of pink yarn. How many balls of yarn in all?

5. Henry had 7 candies and gave 4 away. How many had he left?

6. Levi made 5 paper airplanes in the morning and 9 in the evening. How many in all?

7. Silas had 15 dollars. He paid the shopkeeper $5. How much had he left?

8. 2 kittens in a basket. How many legs and tails in all?

9. There were 6 cups on a shelf. 2 cups fell and broke. How many remained?

10. Elianna read 5 chapters of her book in the morning and 8 in the evening. How many chapters did she read in all?

11. 5 + 3 =

12. 12 – 2 =

13. 12 – 3 =

14. 2 + 6 =

15. 2 + 9 =

16. 5 + 7 =

17. 13 – 5 =

18. 4 + 7 =

19. 9 – 3 =

20. 4 – 3 =

Answers

1. 7 roses
2. 5 robins
3. 3 wheels
4. 11 balls of yarn
5. 3 candies
6. 14 paper airplanes
7. $10
8. 10 legs and tails
9. 4 cups
10. 13 chapters
11. 8
12. 10
13. 9
14. 8
15. 11
16. 12
17. 8
18. 11
19. 6
20. 1

Addition & Subtraction with 6 and Previous Tables

1. There were 9 eggs in a nesting box. If 3 were taken out, how many remain in the box?

2. There were 5 books on a table and 4 on a chair. How many books in all?

3. Peter read 6 pages of his book and then read another 8 pages. How many had he read all together?

4. There were 11 roses on a bush and Jeanne picked 3. How many roses were left on the bush?

5. Mrs. Logan needs to make 10 sandwiches. She has made 6. How many has she left to make?

6. There are 4 balls of pink yarn and 3 balls of orange. How many in all?

7. We have 2 cups of flour in our bowl. How many more must we add to make 6?

8. 15 leaves on a branch when 6 were blown off. How many leaves are left?

9. Mabel is 6 years old and her sister is 12. How many years older than Mabel is her sister?

10. If you have 4 beads, how many more are needed to make 13?

11. 6 + 5 =

12. 12 – 3 =

13. 2 + 5 =

14. 13 – 3 =

15. 16 – 6 =

16. 14 – 6 =

17. 3 + 7 =

18. 4 + 7 =

19. 2 + 8 =

20. 5 + 8 =

Answers

1. 6 eggs
2. 9 books
3. 14 pages
4. 8 roses
5. 4 sandwiches
6. 7 balls of yarn
7. 4 cups
8. 9 leaves
9. 6 years
10. 9 beads
11. 11
12. 9
13. 7
14. 10
15. 10
16. 8
17. 10
18. 11
19. 10
20. 13

Addition & Subtraction with 7 and Previous Tables

1. If there are 13 grapes in a bunch and you eat 7 of them, how many remain?

2. Smith and Bo each had 7¢. How much had they in all?

3. If you had 12¢ but lost 5¢, how much would you have left?

4. There were 3 children skating when 6 more joined them. How many children are now skating?

5. If 4 of those skating children had to go home, how many remain on the rink? (9 – 4)

6. Baron must peel 14 apples. He has peeled 7. How many remain to be peeled?

7. Karen blew up 6 red balloons and 5 blue balloons. How many balloons in all?

8. If 2 of those balloons popped, how many remain? (11 – 2)

9. Can you tell me two numbers that together make 6?

10. Can you tell me two more numbers that together make 6?

11. 2 + 7 =

12. 3 + 3 =

13. 6 + 10 =

14. 14 – 6 =

15. 12 – 3 =

16. 5 + 4 =

17. 8 – 7 =

18. 5 – 2 =

19. 7 + 8 =

20. 6 + 6 =

Answers

1. 6 grapes
2. 14¢
3. 7¢
4. 9 children
5. 5 children
6. 7 apples
7. 11 balloons
8. 9 balloons
9. Any of these: 0 and 6, 1 and 5, 2 and 4, 3 and 3, 4 and 2, 5 and 1, 6 and 0
10. Any of these: 0 and 6, 1 and 5, 2 and 4, 3 and 3, 4 and 2, 5 and 1, 6 and 0
11. 9
12. 6
13. 16
14. 8
15. 9
16. 9
17. 1
18. 3
19. 15
20. 12

Addition & Subtraction with 8 and Previous Tables

1. We borrowed 12 books from the library and returned 8. How many library books do we still have?

2. If we check out 6 more books in addition to the 4 already at home, how many library books will we then have?

3. Caiden had 10 beads but lost 5. How many beads has he now?

4. There are 6 white eggs and 3 brown eggs in the basket. How many eggs in all?

5. 8 penguins on the ice and 5 in the water. How many penguins in all?

6. Drew paid $3 for a pen and $4 for a notepad. How much did he spend in all?

7. Morgan paid $5 for a movie ticket and $4 for a bag of popcorn. How much did she spend all together?

8. Jenny harvested 8 carrots but needs 14. How many more must she pull up?

9. There were 7 cars in the parking lot. If 3 more cars parked, how many in all?

10. Charlotte is 9 and her sister is 5. How many years older is Charlotte?

11. 8 + 8 =

12. 15 – 5 =

13. 12 – 3 =

14. 3 + 5 =

15. 2 + 9 =

16. 11 – 2 =

17. 4 + 4 =

18. 8 + 9 =

19. 11 – 4 =

20. 11 – 3 =

Answers

1. 4 books
2. 10 books
3. 5 beads
4. 9 eggs
5. 13 penguins
6. $7
7. $9
8. 6 carrots
9. 10 cars
10. 4 years
11. 16
12. 10
13. 9
14. 8
15. 11
16. 9
17. 8
18. 17
19. 7
20. 8

Addition & Subtraction with 9 and Previous Tables

1. Jo gave the shopkeeper 10¢ for a 4¢ candy. How much change should she receive?

2. Wyatt and Silas each bought a 5¢ gum ball. How much did they spend in all?

3. Silas had 9¢ and his mother gave him 6¢ more. How much had he all together?

4. Hudson took 7¢ out of his piggy bank but needs 15¢. How much more should he take out?

5. Bridgett gave 6¢ for an eraser and 8¢ for a pencil. How much did she spend?

6. Mimi gave the shopkeeper 10¢ for a candy that cost 8¢. How much change should she get back?

7. Max wants a toy that costs 11¢. He has counted out 9¢. How much more does he need?

8. Levi had 6¢ and his sister gave him 6¢ more. How much has Levi?

9. Aviyah had 13¢ then gave 9¢ for a sticker. How much had she remaining?

10. Anthony has 5¢ more than his brother. If he has 11¢, how much has his brother?

11. 9 + 7 =

12. 16 – 7 =

13. 3 + 9 =

14. 6 + 5 =

15. 2 + 8 =

16. 14 – 4 =

17. 10 – 3 =

18. 9 + 8 =

19. 17 – 9 =

20. 12 – 8 =

Answers

1. 6¢
2. 10¢
3. 15¢
4. 8¢
5. 14¢
6. 2¢
7. 2¢
8. 12¢
9. 4¢
10. 6¢
11. 16
12. 9
13. 12
14. 11
15. 10
16. 10
17. 7
18. 17

19. 8
20. 4

Addition & Subtraction with 10 and Previous Tables

1. Fiona had 8 candles on her cake. She blew out 5. How many are still lit?

2. Milly and Molly each collected 8 seashells. How many had they in all?

3. Mark needs 12 pears for his recipe. He only has 9. How many more does he need?

4. There are 9 cows outside and 4 in the barn. How many cows in all?

5. Seth wants to buy two 10¢ candies. How much does he need?

6. Jared had 10 muffins. He handed out 7. How many had he remaining?

7. Leo needs to save $18. He has saved $10. How much more must he save?

8. You had 17 balloons. If 9 popped, how many remain?

9. If you blew up 6 more balloons, how many would you then have? (8 + 6)

10. If 7 of those then popped, how many balloons would you then have? (14 – 7)

11. 9 + 9 =

12. 14 – 10 =

13. 14 – 9 =

14. 3 + 5 =

15. 8 + 4 =

16. 2 + 7 =

17. 4 – 1 =

18. 10 – 2 =

19. 15 – 6 =

20. 7 – 3 =

Answers

1. 3 candles
2. 16 seashells
3. 3 pears
4. 13 cows
5. 20¢
6. 3 muffins
7. $8
8. 8 balloons
9. 14 balloons
10. 7 balloons
11. 18
12. 4
13. 5
14. 8
15. 12
16. 9

17. 3
18. 8
19. 9
20. 4

Mixed Simple Addition & Subtraction

1. How many ears have 2 boys all together?

2. How many legs have a boy and his cat?

3. How many noses have a girl and her dog?

4. How many tails have 2 cats and 3 dogs?

5. Jenny had 6 eggs and broke 3. How many has she left?

6. Corin baked 10 muffins and 2 were eaten. How many muffins remain?

7. Wyatt planted 9 flowers of which only 5 bloomed. How many did not bloom?

8. Silas has 5 pockets on his pants and 2 on his shirt. How many pockets in all?

9. Sarah did 3 somersaults and then did 3 more. How many somersaults did she do in all?

10. John prepared 6 bowls of popcorn and 4 cups of apple cider for the movie. How many items did he prepare in all?

11. There are 9 bats in a cave and 7 in the next cave. How many bats in all?

12. Shepherd had 8 pennies in his pocket but lost 3. How many pennies had he left?

13. Evienne had 7 balloons and her sister gave her 2 more. How many balloons had she total?

14. There were 12 stamps in a book and 2 were used. How many stamps were left?

15. Abby scooped 7 scoops of chocolate ice cream and 6 scoops of vanilla. How many scoops in all?

16. 8 birds on a wire when 4 flew away. How many birds remained?

17. Peter checked out 10 books but returned 5. How many books has he left?

18. Lily read 7 chapters and then read 3 more. How many chapters did she read in all?

19. How many tails have 6 horses and 6 ponies?

20. How many legs in all have 2 boys and a dog?

Answers

1. 4 ears
2. 6 legs
3. 2 noses
4. 5 tails
5. 3 eggs
6. 8 muffins
7. 4 flowers
8. 7 pockets
9. 6 somersaults
10. 10 items
11. 16 bats

12. 5 pennies
13. 9 balloons
14. 10 stamps
15. 13 scoops
16. 4 birds
17. 5 books
18. 10 chapters
19. 12 tails
20. 8 legs

Mixed Addition & Subtraction Money Sums

1. 4¢ + 4¢ =
2. 6¢ + 3¢ =
3. 8¢ – 2¢ =
4. 8¢ – 4¢ =
5. 3¢ + 9¢ =
6. 7¢ – 4¢ =
7. 5¢ + 6¢ =
8. 8¢ + 8¢ =
9. 4¢ – 1¢ =
10. 2¢ + 9¢ =
11. 5¢ – 1¢ =

12. 18¢ – 10¢ =

13. 12¢ – 5¢ =

14. 5¢ + 2¢ =

15. 1¢ + 8¢ =

16. 7¢ + 10¢ =

17. 12¢ – 6¢ =

18. 7¢ + 9¢ =

19. 4¢ + 8¢ =

20. 8¢ + 3¢ =

Answers

1. 8¢
2. 9¢
3. 6¢
4. 4¢
5. 12¢
6. 3¢
7. 11¢
8. 16¢
9. 3¢
10. 11¢
11. 4¢
12. 8¢
13. 7¢
14. 7¢
15. 9¢
16. 17¢

17. 6¢
18. 16¢
19. 12¢
20. 11¢

Mixed Addition & Subtraction with 3 or More Figures

1. A boy had 10¢ and gave 4¢ to his brother and 4¢ to his sister. How much had he remaining?

2. I had 10¢ and spent 5¢ on a gum ball and then found 2¢ more in my pocket. How much had I?

3. Peter had 6¢, Lily had 4¢, and Stuart had 5¢. How much had they in all?

4. I had 8¢ and gave half to my friend and then spent 1¢ for a gum ball. How much had I remaining?

5. I had 10¢ and lost 6¢ but then found 5¢. How much had I?

6. Ken had 10¢ and bought an eraser for 2¢, a ball for 5¢, and a key for 3¢. Had he anything left?

7. Fiona had 8¢ in her piggy bank, 4¢ in her pocket, and 5¢ on her desk. How much had she in all?

8. I had 15¢ and gave 5¢ for a fairy sticker, 3¢ for a rainbow sticker, and 5¢ for a unicorn sticker. Had I any money left?

9. A boy had 20¢. He gave 5¢ for a candy cane, 5¢ for a gum ball, and 10¢ for a chocolate. Had he any money left?

10. Oleg had 6¢ in his left jeans pocket, 5¢ in his right jeans pocket, and 7¢ in his jacket pocket. How much had he in all?

Answers

1. 2¢
2. 7¢
3. 15¢
4. 3¢
5. 9¢
6. No
7. 17¢
8. Yes. 2¢
9. No
10. 18¢

Mixed Addition & Subtraction in Pure Number

1. 6 + 6 =

2. 6 + 8 =

3. 2 + 7 =

4. 9 – 4 =

5. 7 – 7 =

6. 4 + 8 =

7. 5 + 3 =

8. 3 + 9 =

9. 12 – 2 =

10. 8 – 3 =

11. 4 + 5 =

12. 1 + 10 =

13. 5 + 7 =

14. 7 + 4 =

15. 10 – 3 =

16. 7 – 5 =

17. 10 – 6 =

18. 3 – 2 =

19. 4 + 9 =

20. 20 – 10 =

Answers

1. 12
2. 14
3. 9
4. 5
5. 0
6. 12
7. 8
8. 12
9. 10
10. 5
11. 9

12. 11
13. 12
14. 11
15. 7
16. 2
17. 4
18. 1
19. 13
20. 10

Combined Addition & Subtraction with 3 or More Figures

1. 5 + 1 – 3 =

2. 2 + 4 – 1 =

3. 6 – 5 + 2 =

4. 8 + 2 – 7 =

5. 9 + 4 – 3 =

6. 4 – 4 + 8 =

7. 11 + 3 – 2 =

8. 15 – 10 + 5 =

9. 3 + 3 + 3 – 1 =

10. 12 + 2 – 3 + 4 =

11. 7 – 5 + 8 – 6 =

12. $6 + 6 + 5 - 3 =$

13. $9 - 3 - 2 + 4 =$

14. $8 - 7 + 6 - 2 =$

15. $3 + 2 + 1 - 5 + 4 =$

16. $4 + 4 - 2 + 3 - 5 =$

17. $7 + 5 + 4 - 2 - 3 =$

18. $6 - 5 + 8 - 3 + 2 =$

19. $5 + 6 + 7 - 4 - 4 =$

20. $3 + 8 - 6 + 5 + 7 =$

Answers

1. 3
2. 5
3. 3
4. 3
5. 10
6. 8
7. 12
8. 10
9. 8
10. 15
11. 4
12. 14
13. 8
14. 5
15. 5
16. 4

17. 11
18. 8
19. 10
20. 17

Combined Addition & Subtraction with 5 or More Figures

1. 5 + 2 – 3 – 2 + 6 =

2. 10 – 4 + 7 – 5 + 8 =

3. 13 + 3 + 4 – 2 – 6 =

4. 8 – 4 + 8 + 9 – 7 =

5. 12 – 6 + 10 – 3 + 4 =

6. 7 + 8 + 5 – 2 – 2 =

7. 9 – 7 + 9 + 8 – 5 =

8. 7 + 7 + 3 – 6 + 2 =

9. 10 + 10 + 9 – 3 – 4 =

10. 8 + 6 + 5 – 6 – 3 =

11. 4 + 3 + 2 + 1 – 7 – 3 =

12. 15 – 4 + 6 – 2 + 5 – 3 =

13. Half of 12, + 2 – 5 + 10 + 4 + 4 =

14. 6 + 7 + 4 – 3 – 1 + 5 =

15. 9 + 6 + 2 – 4 + 3 – 8 =

16. 8 + 8 – 6 + 4 – 7 + 3 =

17. 8 – 5 – 3 + 15 – 4 + 10 =

18. Half of 20, + 9 – 5 + 4 – 7 + 2 =

19. 10 – 8 + 3 + 3 – 4 + 7 =

20. 17 + 3 + 5 – 2 + 7 – 2 – 4 =

Answers

1. 8
2. 16
3. 12
4. 14
5. 17
6. 16
7. 14
8. 13
9. 22
10. 10
11. 0
12. 17
13. 21
14. 18
15. 8
16. 10
17. 21
18. 13
19. 11
20. 24

Multiplication & Division with 3 and Previous Tables

1. 3 hens laid 5 eggs each last week. How many total eggs were laid?

2. How many legs have 3 horses?

3. How many legs have 2 horses?

4. If a stool has 3 legs, how many legs have 7 stools?

5. There are 8 apples to share equally among 2 boys. How many apples does each boy get?

6. If gum drops cost 3¢ each, how many can be had for 18¢?

7. If you can walk 3 miles in an hour, how many miles can you walk in 3 hours?

8. There are 7 children. How many legs in all?

9. There are 3 kittens in a box. How many ears have they in all?

10. How many legs?

11. How many 3s in 24?

12. 2 taken 5 times equals?

13. How many 2s in 20?

14. How many 3s in 15?

15. 3 times 2 equals?

16. 3 taken 9 times gives?

17. How many 2s in 12?

18. How many 2s in 16?

19. What is 3 taken 7 times?

20. How many 3s in 27?

Answers

1. 15 eggs
2. 12 legs
3. 8 legs
4. 21 legs
5. 4 apples
6. 6 gum drops
7. 9 miles
8. 14 legs
9. 6 ears
10. 12 legs
11. 8
12. 10
13. 10
14. 5
15. 6
16. 27
17. 6
18. 8
19. 21
20. 9

Multiplication & Division with 4 and Previous Tables

1. Cole picked 2 peaches at a time. If he picked 5 times, how many peaches?

2. How many legs have 7 lambs?

3. How many legs have 9 lambs?

4. Andy paid $16 dollars for 4 loaves of bread. How much was each loaf?

5. There are 32 cards to be divided equally among 4 children. How many cards does each child receive?

6. If each bag holds 4 books, how many books are held in 6 bags?

7. How many arms have 3 children?

8. How many horses can be shod with 20 horseshoes?

9. Each basket holds 3 balls of yarn. How many balls of yarn in 4 baskets?

10. At 4¢ apiece, how many gum balls can be had for 12¢?

11. How many 3s in 15?

12. How many 2s in 14?

13. 2 multiplied by 6 equals?

14. 4 taken 5 times gives?

15. How many 4s in 36?

16. How many 3s in 21?

17. How much is 3 taken 6 times?

18. How much is 2 taken 8 times?

19. How much is 4 taken 6 times?

20. How many 3s in 9?

Answers

1. 10 peaches
2. 28 legs
3. 36 legs
4. $4
5. 8 cards
6. 24 books
7. 6 arms
8. 5 horses
9. 12 balls of yarn
10. 3 gum balls
11. 5
12. 7
13. 12
14. 20
15. 9
16. 7
17. 18
18. 16
19. 24
20. 3

Multiplication & Division with 5 and Previous Tables

1. One nickel is 5¢. How much is 5 nickels?

2. How many cents total are 7 nickels?

3. If a sticker costs 2¢, how many can be had for 20¢?

4. If a ribbon costs 5¢, how many can be had for 35¢?

5. If you bicycle 4 miles an hour, how far will you go in 4 hours?

6. How far will you travel in 8 hours? (4 x 8)

7. 12 ears of sweet corn to be shared equally among 4 people. How many ears does each get?

8. 2 cats and 2 kittens. How many legs among them?

9. If 7 children are wearing shoes, how many shoes in all?

10. How many shoes have 7 horses?

11. How many 5s in 45?

12. How many 3s in 24?

13. How much is 5 taken 4 times?

14. How much is 2 taken 8 times?

15. How much is 5 multiplied by 6?

16. How much is 3 times 6?

17. How many 5s in 40?

18. How many 4s in 40?

19. How much is 2 taken 3 times?

20. How much is 3 taken 9 times?

Answers

1. 25¢
2. 35¢
3. 10 stickers
4. 7 ribbons
5. 16 miles
6. 32 miles
7. 3 ears
8. 16 legs
9. 14 shoes
10. 28 shoes
11. 9
12. 8
13. 20
14. 16
15. 30
16. 18
17. 8
18. 10
19. 6
20. 27

Multiplication & Division with 6 and Previous Tables

1. If honey sticks are 5¢ each, how many can be bought with 25¢?
2. If 18 sheep are divided equally into 3 flocks, how many sheep in each flock?
3. Ben can bring in 3 bags of groceries each trip. How many bags in 3 trips?
4. There are 4 tires on a car. How many tires on 6 cars?
5. How many tires on 9 cars?
6. The Smiths hike 5 miles in an hour. How far do they hike in 3 hours?
7. A yard is 3 feet long. How many feet are in 2 yards?
8. How many yards in 9 feet?
9. There are 6 slices in a pizza. How many slices in 6 pizzas?
10. And if there are 24 slices of pizza, how many pizzas are there?
11. How many 5s in 35?
12. How many 6s in 18?
13. 4 times 3 equals?
14. 2 times 7 equals?
15. 6 taken 9 times equals?
16. How many 3s in 21?

17. How many 5s in 45?

18. How much is 3 taken 9 times?

19. How much is 4 taken 4 times?

20. How many 6s in 48?

Answers

1. 5 honey sticks
2. 6 sheep
3. 9 bags
4. 24 tires
5. 36 tires
6. 15 miles
7. 6 feet
8. 3 yards
9. 36 slices
10. 4 pizzas
11. 7
12. 3
13. 12
14. 14
15. 54
16. 7
17. 9
18. 27
19. 16
20. 8

Mixed Multiplication & Division

1. At 3¢ each, how many bracelets can I buy for 15¢?

2. Jeffrey walks 2 miles each day. How many miles does he walk in 4 days?

3. At 2 dimes apiece, how many trading cards can I buy for 8 dimes?

4. There are 3 feet in a yard. How many yards are in 12 feet?

5. 16 marbles divided evenly between 2 boys gives each boy how many marbles?

6. How many candies at 4¢ apiece can we purchase for 36¢?

7. There are 5 sunflowers planted in a row. How many in 6 rows?

8. How far will a nuthatch travel at a rate of 4 miles a day over 6 days?

9. At 2¢ each, how many gummy bears can I buy for 18¢?

10. Our family hiked a total of 6 miles in 2 hours. How many miles had we hiked in 1 hour?

11. How many orange slices at 2¢ each can be bought for 12¢?

12. 4 kittens have how many legs in all?

13. 4 kittens have how many eyes in all?

14. How many yards of lace at $1 a yard can be bought for $9?

15. How many ribbons at 5¢ each can be bought for 30¢?

16. A man gave $50 for some bundles of firewood. If each bundle cost $5, how many bundles did he buy?

17. How many walnuts can I buy for 24¢ if each costs 4¢?

18. Jeremy earns $6 an hour for picking up crabapples. How many hours must he work to earn $36?

19. 15¢ will buy how many gum drops at 5¢ each?

20. A boy and 2 dogs have how many tails between them?

Answers

1. 5 bracelets
2. 8 miles
3. 4 trading cards
4. 4 yards
5. 8 marbles
6. 9 candies
7. 30 sunflowers
8. 24 miles
9. 9 gummy bears
10. 3 miles
11. 6 orange slices
12. 16 legs
13. 8 eyes
14. 9 yards
15. 6 ribbons
16. 10 bundles
17. 6 walnuts
18. 6 hours
19. 3 gum drops
20. 2 tails

Multiplication with Money Problems

Find the cost of

1. 3 postcards at 9¢ each.
2. 10 goldfish at 3¢ each.
3. 5 stickers at 5¢ each.
4. 6 candies at 4¢ each.
5. 4 pencils at 3¢ each.
6. 10 gumballs at 2¢ each.
7. 6 bracelets at 6¢ each.
8. 3 marbles at 3¢ each.
9. 9 hairbands at 4¢ each.
10. 8 flowers at 2¢ each.

Answers

1. 27¢
2. 30¢
3. 25¢
4. 24¢
5. 12¢
6. 20¢
7. 36¢
8. 9¢
9. 36¢
10. 16¢

Division with Money Problems

How many

1. 5¢ stickers for 15¢?
2. 4¢ pencils for 20¢?
3. 2¢ gum balls for 16¢?
4. 6¢ candies for 36¢?
5. 3¢ toys for 24¢?
6. 1¢ stamps for 6¢?
7. 5¢ oranges for 45¢?
8. 5¢ apples for 30¢?
9. 4¢ erasers for 12¢?
10. 6¢ lollipops for 42¢?

Answers

1. 3 stickers
2. 5 pencils
3. 8 gum balls
4. 6 candies
5. 8 toys
6. 6 stamps
7. 9 oranges
8. 6 apples
9. 3 erasers
10. 7 lollipops

Multiplication in Pure Number

1. $4 \times 3 =$
2. $2 \times 8 =$
3. $1 \times 6 =$
4. $3 \times 7 =$
5. $5 \times 2 =$
6. $6 \times 8 =$
7. $3 \times 9 =$
8. $2 \times 7 =$
9. $4 \times 4 =$
10. $6 \times 2 =$
11. $5 \times 5 =$
12. $6 \times 6 =$
13. $2 \times 9 =$
14. $1 \times 6 =$
15. $3 \times 5 =$
16. $4 \times 9 =$
17. $6 \times 3 =$
18. $4 \times 7 =$

19. $2 \times 8 =$

20. $4 \times 2 =$

21. $6 \times 5 =$

22. $5 \times 7 =$

23. $3 \times 8 =$

24. $2 \times 7 =$

25. $4 \times 8 =$

26. $6 \times 9 =$

27. $4 \times 10 =$

28. $5 \times 8 =$

29. $2 \times 2 =$

30. $3 \times 2 =$

Answers

1. 12
2. 16
3. 6
4. 21
5. 10
6. 48
7. 27
8. 14
9. 16
10. 12
11. 25

12. 36
13. 18
14. 6
15. 15
16. 36
17. 18
18. 28
19. 16
20. 8
21. 30
22. 35
23. 24
24. 14
25. 32
26. 54
27. 40
28. 40
29. 4
30. 6

Division in Pure Number

1. 24 ÷ 3 =

2. 6 ÷ 2 =

3. 12 ÷ 3 =

4. 25 ÷ 5 =

5. 12 ÷ 2 =

6. 16 ÷ 4 =

7. 30 ÷ 3 =

8. 10 ÷ 2 =

9. 20 ÷ 4 =

10. 10 ÷ 5 =

11. 35 ÷ 5 =

12. 12 ÷ 4 =

13. 27 ÷ 3 =

14. 48 ÷ 6 =

15. 36 ÷ 4 =

16. 21 ÷ 3 =

17. 16 ÷ 2 =

18. 6 ÷ 2 =

19. 42 ÷ 6 =

20. 9 ÷ 3=

21. 20 ÷ 5 =

22. 18 ÷ 2 =

23. 28 ÷ 4 =

24. 45 ÷ 5 =

25. 32 ÷ 4 =

26. 15 ÷ 3 =

27. 20 ÷ 2 =

28. 60 ÷ 6 =

29. 40 ÷ 5 =

30. 24 ÷ 6 =

Answers

1. 8
2. 3
3. 4
4. 5
5. 6
6. 4
7. 10
8. 5
9. 5
10. 2
11. 7
12. 3
13. 9
14. 8
15. 9
16. 7
17. 8
18. 3
19. 7
20. 3
21. 4
22. 9
23. 7

24. 9
25. 8
26. 5
27. 10
28. 10
29. 8
30. 4

Combined Multiplication and Division in Pure Number

1. $3 \times 4 \div 6 =$

2. $12 \div 2 \times 4 =$

3. $5 \times 4 \div 2 =$

4. $4 \times 4 \div 2 =$

5. $3 \times 6 \div 2 =$

6. $4 \times 9 \div 6 =$

7. $6 \times 4 \div 3 =$

8. $6 \times 2 \div 4 =$

9. $5 \times 8 \div 4 =$

10. $4 \times 7 \div 2 =$

Answers

1. 2
2. 24

3. 10
4. 8
5. 9
6. 6
7. 8
8. 3
9. 10
10. 14

Exams

This book contains three term exams for your convenience. Exams traditionally take place in the final week of a 12-week term (or your chosen divisions of your school year). Through daily work you will already have a good understanding of your child's progress, so these exams serve mainly as a record of what your child knows rather than a test of what your child does not know. Adjust questions to reflect what your child has accomplished each term.

Exam questions are mainly oral with some reading and writing. Answers may be recorded by the parent or by the student, depending on the student's ability. Concrete objects or manipulatives may be used if your child is in the concrete stage. Only take as many questions as your child is able to answer in 15–20 minutes.

Term I Exam

Concrete objects may be used.

Tip: Change the numbers based on where your child is in the lessons.

1. William gathered 5 pounds of potatoes in the morning and 8 pounds of potatoes in the evening. How many pounds of potatoes had he gathered in all?

2. There are 13 chickens in the coop and 4 go outside. How many chickens remain in the coop?

3. Sarai wants to buy a pencil that costs 15¢. She has only 7¢. How much more does she need?

4. 9 - 5 =

5. 10 + 10 =

6. 8 + 4 =

7. 6 – 3 =

8. Asher records the birds he sees in his nature journal. He observed 5 robins, 4 blue jays, and 6 starlings in the morning and 4 hummingbirds and 2 doves in the afternoon. How many total birds did Asher record?

Answers

1. 13 pounds
2. 9 chickens
3. 8¢
4. 4
5. 20
6. 12
7. 3
8. 21 birds

Term II Exam

Concrete objects may be used.

Tip: Change the numbers based on where your child is in the lessons.

1. How many units in 546? How many tens? How many hundreds?

2. How many units in 400? How many tens? How many hundreds?

3. Read these numbers: 274, 638, 942.

4. Write these numbers: twenty-three, ninety-six, two hundred two, seven hundred eighty-one, one thousand.

5. $\begin{array}{r} 284 \\ +563 \\ \hline \end{array}$ 6. $\begin{array}{r} 159 \\ +761 \\ \hline \end{array}$ 7. $\begin{array}{r} 369 \\ -158 \\ \hline \end{array}$ 8. $\begin{array}{r} 643 \\ -587 \\ \hline \end{array}$

Answers

1. 6 units, 4 tens, 5 hundreds
2. 0 units, 0 tens, 4 hundreds
3. Two hundred seventy-four, six hundred thirty-eight, nine hundred forty-two
4. 23, 96, 202, 781, 1000

5. $\begin{array}{r} {\scriptstyle 1\ } \\ 284 \\ +563 \\ \hline 847 \end{array}$ 6. $\begin{array}{r} {\scriptstyle 1\,1\ } \\ 159 \\ +761 \\ \hline 920 \end{array}$ 7. $\begin{array}{r} 369 \\ -158 \\ \hline 211 \end{array}$ 8. $\begin{array}{r} {\scriptstyle 5\;13\,1} \\ \not{6}\not{4}3 \\ -587 \\ \hline 56 \end{array}$

Term III Exam

Concrete objects may be used.

Tip: Change the numbers based on where your child is in the lessons.

1. If Lily makes 3 pancakes for each member of her family, how many pancakes must she make for 6 people?

2. The orchard has 4 rows of peach trees with 5 trees in each row. How many peach trees in all?

3. What is 4 x 1000?

4. What is 6 x 100?

5. How many 6s in 42?

6. How many 3s in 24?

7. How many candies at 5¢ each can you buy for 40¢?

8. How many rubber balls at 4¢ each can be had for 28¢? Write this as a division problem.

Answers

1. 18 pancakes
2. 20 peach trees
3. 4000
4. 600
5. 7
6. 8
7. 8 candies
8. 7 rubber balls. Written as 28 ÷ 4 = 7 or $4\overline{)28}$ with 7 above

Once your student has completed these lessons, he is ready to move on to Book 3 in *The Charlotte Mason Elementary Arithmetic Series*. In Book 3 your student will continue to master the multiplication tables up to 10 x 10, begin to work with long division, receive a formal introduction to simple fractions, and learn to work with numbers up to 10,000. He will also enjoy a gentle introduction to weights and measures and explore the concept of time.